S0-HWF-560
R0419195

WATER RESOURCE PLANNING, DEVELOPMENT AND MANAGEMENT

DRINKING WATER AND UNREGULATED CONTAMINANTS

ANALYSES OF EPA MEASURES

WATER RESOURCE PLANNING, DEVELOPMENT AND MANAGEMENT

Additional books in this series can be found on Nova's website under the Series tab.

Additional e-books in this series can be found on Nova's website under the e-book tab.

WATER RESOURCE PLANNING, DEVELOPMENT AND MANAGEMENT

DRINKING WATER AND UNREGULATED CONTAMINANTS

ANALYSES OF EPA MEASURES

CYNTHIA ROMERO
EDITOR

nova publishers
New York

Copyright © 2014 by Nova Science Publishers, Inc.

All rights reserved. No part of this book may be reproduced, stored in a retrieval system or transmitted in any form or by any means: electronic, electrostatic, magnetic, tape, mechanical photocopying, recording or otherwise without the written permission of the Publisher.

For permission to use material from this book please contact us:
Telephone 631-231-7269; Fax 631-231-8175
Web Site: http://www.novapublishers.com

NOTICE TO THE READER

The Publisher has taken reasonable care in the preparation of this book, but makes no expressed or implied warranty of any kind and assumes no responsibility for any errors or omissions. No liability is assumed for incidental or consequential damages in connection with or arising out of information contained in this book. The Publisher shall not be liable for any special, consequential, or exemplary damages resulting, in whole or in part, from the readers' use of, or reliance upon, this material. Any parts of this book based on government reports are so indicated and copyright is claimed for those parts to the extent applicable to compilations of such works.

Independent verification should be sought for any data, advice or recommendations contained in this book. In addition, no responsibility is assumed by the publisher for any injury and/or damage to persons or property arising from any methods, products, instructions, ideas or otherwise contained in this publication.

This publication is designed to provide accurate and authoritative information with regard to the subject matter covered herein. It is sold with the clear understanding that the Publisher is not engaged in rendering legal or any other professional services. If legal or any other expert assistance is required, the services of a competent person should be sought. FROM A DECLARATION OF PARTICIPANTS JOINTLY ADOPTED BY A COMMITTEE OF THE AMERICAN BAR ASSOCIATION AND A COMMITTEE OF PUBLISHERS.

Additional color graphics may be available in the e-book version of this book.

Library of Congress Cataloging-in-Publication Data

ISBN: 978-1-63321-164-3

Published by Nova Science Publishers, Inc. † New York

CONTENTS

PREFACE

The Environmental Protection Agency (EPA) has implemented all of the recommendations GAO made in its May 2011 report to improve the Unregulated Contaminant Monitoring Rule (UCMR) program. EPA's UCMR program collects data on unregulated contaminants in the nation's drinking water. EPA uses this data and other information to make decisions on whether to regulate additional drinking water contaminants. This book examines the extent to which EPA implemented GAO's prior recommendations to improve the program and opportunities, if any, to strengthen it further; the factors EPA considered when it selected the UCMR3 contaminants and the limitations, if any, it faced in selecting them; and the extent to which UCMR data support regulatory determinations. It also discusses and summarizes the Safe Drinking Water Act (SDWA) and its major programs and regulatory requirements.

Chapter 1 – EPA's UCMR program collects data on unregulated contaminants in the nation's drinking water. EPA uses these data and other information to make regulatory determinations— decisions on whether to regulate additional drinking water contaminants. It is currently in its third data collection cycle, UCMR3.

GAO was asked to examine the UCMR program. This report examines: (1) the extent to which EPA implemented GAO's prior recommendations to improve the program and opportunities, if any, to strengthen it further; (2) the factors EPA considered when it selected the UCMR3 contaminants and the limitations, if any, it faced in selecting them; and (3) the extent to which UCMR data support regulatory determinations.

GAO reviewed EPA documents, surveyed 48 subject matter experts, assessed the UCMR program against statutory requirements and other standards, and interviewed EPA officials.

Chapter 2 – Under 1996 amendments to the Safe Drinking Water Act, every 5 years the Environmental Protection Agency (EPA) is to determine for at least five contaminants, such as chemicals, whether regulation is warranted, considering those that present the greatest public health concern. Since 1996, EPA had not recommended any new contaminants for regulation until February 2011, when it reversed its controversial 2008 preliminary decision to not regulate perchlorate, an ingredient in rocket fuel and other products. GAO was asked to (1) evaluate the extent to which EPA's implementation of the 1996 amendments has helped assure the public of safe drinking water and (2) review the process and scientific analyses used to develop the 2008 preliminary regulatory determination on perchlorate. GAO analyzed relevant statutory provisions and regulatory determination documents and interviewed EPA officials.

Chapter 3 – This report summarizes the Safe Drinking Water Act (SDWA) and its major programs and regulatory requirements. This report includes the drinking water security provisions added to the SDWA by the Public Health Security and Bioterrorism Preparedness and Response Act of 2002 (P.L. 107-188), and lead reduction provisions as amended by P.L. 111-380. It also outlines amendments made in December 2013 by P.L. 113-64 (H.R. 3588) to explicitly exempt fire hydrants from coverage under the act's lead plumbing restrictions.

The SDWA, Title XIV of the Public Health Service Act, is the key federal law for protecting public water supplies from harmful contaminants. First enacted in 1974 and substantially amended in 1986 and 1996, the act is administered through programs that establish standards and treatment requirements for public water supplies, control underground injection of wastes, finance infrastructure projects, and protect sources of drinking water. The 1974 law established the current federal-state arrangement in which states may be delegated primary implementation and enforcement authority for the drinking water program. The state-administered Public Water Supply Supervision (PWSS) Program remains the basic program for regulating the nation's public water systems, and 49 states have assumed this authority.

The last major reauthorization of the act was done through the Safe Drinking Water Act Amendments of 1996 (P.L. 104-182), which generally authorized appropriations for SDWA programs through FY2003. As with other EPA-administered statutes having expired funding authority, Congress has continued to appropriate funds for the ongoing SDWA programs.

In addition to reviewing key programs and requirements of the SDWA, this report includes statistics on the number and types of regulated public water systems. It also provides a table that lists all major amendments, with the year of enactment and public law number, and that cross-reference sections of the act with the major *U.S. Code* sections of the codified statute.

In: Drinking Water and Unregulated Contaminants
Editor: Cynthia Romero
ISBN: 978-1-63321-164-3
© 2014 Nova Science Publishers, Inc.

Chapter 1

DRINKING WATER: EPA HAS IMPROVED ITS UNREGULATED CONTAMINANT MONITORING PROGRAM, BUT ADDITIONAL ACTION IS NEEDED*

United States Government Accountability Office

WHY GAO DID THIS STUDY

EPA's UCMR program collects data on unregulated contaminants in the nation's drinking water. EPA uses these data and other information to make regulatory determinations— decisions on whether to regulate additional drinking water contaminants. It is currently in its third data collection cycle, UCMR3.

GAO was asked to examine the UCMR program. This report examines: (1) the extent to which EPA implemented GAO's prior recommendations to improve the program and opportunities, if any, to strengthen it further; (2) the factors EPA considered when it selected the UCMR3 contaminants and the limitations, if any, it faced in selecting them; and (3) the extent to which UCMR data support regulatory determinations.

GAO reviewed EPA documents, surveyed 48 subject matter experts, assessed the UCMR program against statutory requirements and other standards, and interviewed EPA officials.

WHAT GAO RECOMMENDS

Congress should consider amending the Safe Drinking Water Act to allow EPA to monitor for more than 30 contaminants under certain circumstances, and to adjust statutory time frames so UCMR data can inform regulatory determinations in the same cycle. GAO, among other things, recommends that EPA vary the monitoring frequency based on

* This is an edited, reformatted and augmented version of United States Government Accountability Office Publication, No. GAO-14-103, dated January 2014.

contaminant type. In commenting on a draft of this report, EPA generally agreed with GAO's findings, conclusions, and recommendations.

WHAT GAO FOUND

The Environmental Protection Agency (EPA) has implemented all of the recommendations GAO made in its May 2011 report to improve the Unregulated Contaminant Monitoring Rule (UCMR) program. In that report, GAO recommended that EPA (1) monitor for the full 30 contaminants allowed by statute, (2) monitor for most or all contaminants using a more robust monitoring approach, and (3) select sufficiently sensitive minimum reporting levels (MRL) for monitoring contaminants. EPA now requires public water systems to monitor for 30 contaminants in the UCMR3 program, using its most robust monitoring approach for a majority of these contaminants, and setting MRLs as low as can be reliably measured, according to EPA. The Safe Drinking Water Act (SDWA) requires EPA to vary the monitoring frequency based on the type of contaminant likely to be found, but EPA used a standard monitoring frequency for all contaminants. This may result in inaccurate estimates of the occurrence of sporadically occurring microbes (e.g., viruses) or pesticides, according to experts GAO surveyed and studies it reviewed. In such cases, the monitoring data may not provide reliable estimates of contaminant occurrence.

EPA used 10 factors to select the 30 contaminants for UCMR3, but its selection process faced some limitations. Officials told GAO that the contaminants did not have to meet all 10 of the selection factors to be chosen, but 3 were very important (1) the availability of an analytical method to detect contaminants, (2) the reliability of health effects information on the contaminants, and (3) the need for data to support regulatory determinations for priority contaminants. However, EPA is limited by a statutory cap of 30 contaminants every 5 years, which restricts its ability to collect data on additional contaminants that could have been monitored for little additional cost. SDWA's legislative history reflected concerns with the ability of public water systems to absorb such costs, but many of the analytical methods EPA is using for UCMR3 are able to test a single sample of drinking water for more than one contaminant at a time. However, because of the limit of 30, EPA cannot always take advantage of this efficiency and is unable to gain economies of scale using monitoring that is already under way.

EPA uses UCMR data to support regulatory determinations but faces a time lag when doing so. EPA has used UCMR data to support 10 out of 12 regulatory determinations it has made since 2008 and is currently using UCMR data to inform the determinations expected in 2015. However, a time lag between the statutory deadline for making regulatory determinations and when UCMR data are available delays determinations on given contaminants until the following cycle. The 2-year time frame SDWA originally established from the time EPA publishes the UCMR list to when it makes regulatory determinations has not provided enough time for the agency to incorporate the UCMR data into the determinations. The UCMR3 monitoring, data collection, and analysis overlap with the time when EPA will be making its regulatory determinations for contaminants from its most recent Contaminant Candidate List. Consequently, UCMR data are not available to support regulatory determinations for contaminants during the cycle in which they are monitored;

rather, UCMR data typically are not used until the next cycle. EPA officials told GAO that most of the UCMR3 data, which are being collected from 2013 to 2015, will be used to support the regulatory determinations it expects to issue in 2020 instead of 2015.

ABBREVIATIONS

CCL	Contaminant Candidate List
EPA	Environmental Protection Agency
GC-MS	gas chromatograph-mass spectrometer
MRL	minimum reporting level
MTBE	methyl tertiary-butyl ether
NAWQA	National Water-Quality Assessment Program
OMB	Office of Management and Budget
PFC	perfluorinated chemical
SDWA	Safe Drinking Water Act
UCMR	Unregulated Contaminant Monitoring Rule
USGS	United States Geological Survey
VOC	volatile organic compound

January 9, 2014

The Honorable Henry A. Waxman
Ranking Member
Committee on Energy and Commerce
House of Representatives

The Honorable Edward J. Markey
United States Senate

Most Americans receive their drinking water from the more than 160,000 public water systems in the United States, all of which are subject to key requirements of the Safe Drinking Water Act (SDWA).[1] Under SDWA, the Environmental Protection Agency (EPA) is authorized to regulate contaminants in public water systems.[2] Currently, EPA regulates more than 89 drinking water contaminants. In 1996, in the midst of concerns about the number and pace of regulations required under the act and the need to better focus limited resources on contaminants of greatest public health concern, Congress enacted significant amendments to SDWA. The 1996 amendments to SDWA also directed EPA, among other things, to identify unregulated contaminants that present the greatest public health concern, establish a program to monitor drinking water for unregulated contaminants, and decide whether or not to regulate at least 5 such contaminants every 5 years.[3] EPA decisions about whether or not to regulate these contaminants are called regulatory determinations. Since 1996, EPA has made 21 regulatory determinations and has decided to regulate one of the contaminants, perchlorate.[4] EPA has identified 116 unregulated contaminants, including perchlorate and others such as formaldehyde and *Salmonella enterica*, that it believes may occur in public water systems and may also require regulation. Those unregulated contaminants are listed on EPA's current Contaminant Candidate List (CCL).[5]

In addition, EPA has established the Unregulated Contaminant Monitoring Rule (UCMR) program, to monitor drinking water for unregulated contaminants, under the 1996 amendments to SDWA. As part of the UCMR, the SDWA amendments require EPA to issue a list every 5 years identifying up to 30 contaminants to be monitored by public water systems. The UCMR program is currently in its third 5-year cycle, known as UCMR3. EPA estimates that UCMR3 will cost $87 million from 2012 to 2016, of which $21 million will be paid by EPA. EPA will pay for small water systems' costs for the UCMR; the remaining $66 million will be paid by large water systems.[6] The UCMR, CCL, and regulatory determinations are related to each other. First, the CCL identifies unregulated contaminants that may require regulation in the future. Through the UCMR program, public water systems monitor for some of the CCL contaminants to determine how often and at what level the contaminants occur in drinking water. EPA then makes regulatory determinations about contaminants on the CCL using UCMR data, information about the contaminants' health effects, and other information for support.

In May 2011, we reported that EPA's monitoring program under the UCMR did not use EPA's full statutory authority under SDWA, use sufficiently sensitive testing methods, or employ sufficiently robust monitoring approaches.[7] First, EPA did not use its full allotment of 30 contaminants as authorized by SDWA in either of the first two UCMR cycles (UCMR1 and UCMR2), which limited the agency's progress in obtaining data on drinking water contaminants. Second, in some cases, EPA required public water systems to use testing methods (i.e., analytical methods) that were not sufficiently sensitive to identify the presence of contaminants at EPA's health reference level—the health benchmark EPA uses in assessing whether to regulate specific contaminants. Third, EPA used a limited sample size of 300 public water systems for UCMR1 for one of the program's monitoring approaches, which resulted in occurrence data that, according to EPA documents, were not sufficient to make national estimates of contaminant occurrence. In the 2011 report, we also recommended that EPA use its full statutory authority to monitor for 30 unregulated contaminants, select minimum reporting levels that are sufficiently sensitive to detect the presence of contaminants in public water systems at levels of public health concern, and monitor for most or all contaminants using the program's most robust monitoring approach.

In this context, you asked us to examine the UCMR program. Specifically, our objectives for this report were to: (1) evaluate the extent to which EPA has implemented the recommendations we made in 2011 to improve the UCMR program and identify any opportunities to strengthen it further; (2) identify the factors that EPA considered when it selected the contaminants for monitoring under UCMR3 and the limitations, if any, that EPA faced in selecting the contaminants; and (3) examine the extent to which UCMR data support regulatory determinations.

To identify the steps EPA has taken in response to the recommendations we made in 2011, we reviewed EPA documents, including EPA's proposed and final rules for the UCMR program, assessed the UCMR sample design against statutory requirements and statistical standards, and interviewed EPA officials. To identify the factors EPA considers when selecting contaminants, we reviewed EPA documents, including EPA's proposed and final rules for the UCMR program and SDWA requirements, and interviewed EPA officials. To determine how EPA uses the UCMR data to support its regulatory determinations, we reviewed EPA and *Federal Register* documents and interviewed EPA officials. In addition, to obtain further information about the UCMR program and identify opportunities to strengthen it further, we conducted a two-round Delphi survey[8] of experts selected for their knowledge of one or more of the following areas: drinking water, contaminant monitoring, analytical methods, toxicology, risk assessment, and environmental regulation. In the first round of the survey, we selected 14 experts who demonstrated familiarity with EPA's CCL, UCMR, and regulatory determination processes. We asked them 10 open-ended questions about how various aspects of the UCMR program helped or hindered EPA in (1) selecting priority contaminants, (2) monitoring for selected contaminants, and (3) using the data to support regulatory determinations.

We analyzed the responses provided by the experts and developed closed-ended questions for the second round of the survey. To identify experts for the second round of the survey, we developed a preliminary list of experts, which included the 14 experts who completed the first round of the survey and others identified through interviews with stakeholders and our review of public comments provided on EPA's proposed rules.

We contracted with the National Academy of Sciences to identify additional experts. We selected 48 experts for the second round of the survey based on their knowledge of one or more of the following topic areas: drinking water, contaminant monitoring, analytical methods, toxicology, risk assessment, and environmental regulation. For the first round of the survey, we received responses from 14 of the 14 experts we surveyed and, for the second round, we received responses from 40 of 48 experts we surveyed (83 percent response rate). The information that we obtained from the expert surveys is not generalizable to all experts that have an interest in or knowledge of drinking water contaminant monitoring but does provide several experts' views on the UCMR program. For more information on our objectives, scope, and methodology see appendix I. See appendix IV for a copy of the survey questions and responses for the second round of the survey and appendix V for a list of the experts who completed the surveys.

We conducted this performance audit from June 2012 to January 2014 in accordance with generally accepted government auditing standards. Those standards require that we plan and perform the audit to obtain sufficient, appropriate evidence to provide a reasonable basis for our findings and conclusions based on our audit objectives. We believe that the evidence obtained provides a reasonable basis for our findings and conclusions based on our audit objectives.

BACKGROUND

This section presents information on (1) regulated and unregulated contaminants in drinking water, (2) the UCMR program and SDWA requirements, and (3) laboratories and methods used to analyze drinking water.

Regulated and Unregulated Contaminants in Drinking Water

Drinking water contaminants include chemical and microbial substances, some of which can harm human health. For example, prolonged exposure to arsenic in drinking water can cause skin damage or circulatory problems and, over time, may increase cancer risks. Contaminants may be naturally occurring or introduced through other means. EPA has identified three ways that drinking water generally becomes contaminated. First, contaminants can find their way into drinking water sources from industrial waste releases, agricultural runoff, and other pollution sources. Second, contamination can occur during the treatment of water by forming what are known as "disinfection by-products."[9] Third, contaminants can "leach" or dissolve out of the materials used for treatment, storage, and distribution of water.

EPA regulates drinking water contaminants under SDWA by issuing legally enforceable standards, known as National Primary Drinking Water Regulations or "primary standards" that generally limit the levels of these contaminants in public water systems. EPA has issued primary standards addressing more than 89 contaminants. States generally enforce these standards through compliance monitoring. Under compliance monitoring, public water systems collect a sample of drinking water, analyze the sample for specific contaminants, and report the results to states. To analyze the samples, water systems and states use analytical methods developed by government agencies, universities, and other organizations. EPA evaluates analytical methods for drinking water and approves those it determines meet agency requirements for monitoring contaminants. Laboratories analyzing drinking water compliance samples must be certified by the state or EPA. EPA establishes schedules for how often water systems must monitor for regulated contaminants. The monitoring frequency generally depends on the contaminant, previous monitoring results, the type of source water, and the size of the population served by the water system.

Unregulated drinking water contaminants include chemical and microbial substances and are not currently subject to primary standards under SDWA.[10] Exposure to unregulated drinking water contaminants may pose a human health risk, particularly to vulnerable subpopulations such as pregnant women, infants, the elderly, and those with compromised immune systems.[11] Unregulated contaminants include industrial chemicals, military chemicals, pesticides, bacteria, and other microbes and chemicals. They also include contaminants of emerging concern— contaminants characterized by a perceived, potential or real threat to human health or the environment or by a lack of published health standards. Examples include pharmaceuticals, such as prescription and over-the-counter drugs, and ingredients in personal care products, such as cosmetics and soaps. A contaminant may also be considered to be emerging because a new source of public exposure has been discovered or a new detection method or treatment technology has been developed.

UCMR Program and SDWA Requirements

Under the UCMR program, EPA collects data about the occurrence of unregulated contaminants in drinking water. The UCMR program collects data on the extent of the occurrence of unregulated contaminants in all public water systems serving more than 10,000 people and in a nationally representative sample of smaller systems. These data inform EPA's decisions about whether to issue primary standards for unregulated contaminants. Under the

UCMR, EPA is to identify up to 30 unregulated contaminants that may be present in drinking water and establishes a monitoring program to obtain data on the extent of their occurrence. These data help EPA determine whether drinking water contamination is occurring and estimate its level and frequency. EPA established a three-tiered approach for monitoring UCMR contaminants that includes (1) assessment monitoring, (2) screening survey, and (3) pre-screen testing. EPA's approach is based primarily on the availability of analytical methods to monitor the contaminants. Assessment monitoring is used to monitor contaminants that can be detected by well-established analytical methods that are widely used in drinking water laboratories. The screening survey is used to monitor contaminants that can be detected by more recently developed analytical methods that are used in fewer drinking water laboratories. Pre-screen testing is used to monitor contaminants that can be detected by very new or specialized analytical methods that are used in a small set of drinking water laboratories.

Approximately 6,000 public water systems are currently participating in the UCMR program. Participating public water systems are monitoring for contaminants under one or more of the three tiers. EPA varies the number of public water systems required to monitor under each tier. UCMR monitoring is conducted by all large community and nontransient, noncommunity public water systems serving more than 10,000 people.[12] Monitoring is also conducted by a nationally representative sample of other public water systems that EPA selects (see table 1). In addition, a sample of community and nontransient, noncommunity water systems, and transient water systems are required to allow EPA to conduct prescreen testing under UCMR3.[13]

Table 1. Public Water Systems Required to Participate in the Third Unregulated Contaminant Monitoring Rule

Monitoring approach	Type of system	Size of system	Number of systems
Assessment monitoring using well-established analytical methods	Community water systems and nontransient, noncommunity water systems	Systems serving more than 10,000 people	All (~4,200)
Systems serving 10,000 or fewer people		Representative sample of800 systems	
Screening survey using more recently developed analytical methods	Community water systems and nontransient, noncommunity water systems	Systems serving more than 100,000 people	All (~400)
Systems serving 10,001 to 100,000 people		Representative sample of320 systems	
Systems serving 10,000 or fewer people		Representative sample of480 systems	
Pre-screen testing using specialized analytical methods	Nondisinfecting groundwater systems including community water systems, nontransient, noncommunity water systems, and transient, noncommunity water systems	Systems serving 1,000 or fewer people	Representative sample of800 systems

Source: GAO analysis of EPA documents.

The UCMR program is part of a larger framework established by SDWA that also includes the CCL and requires EPA to make regulatory determinations for at least five of the contaminants on that list every 5 years. The CCL is a list of unregulated contaminants that are known or anticipated to occur in public water systems and that may require regulation. SDWA directs EPA to select contaminants that pose the greatest public health concern for consideration for the CCL and regulatory determinations.[14] A regulatory determination is a formal decision on whether EPA should initiate a rulemaking process to develop a primary standard for a specific drinking water contaminant. The UCMR, CCL, and regulatory determinations framework is implemented by three different branches within EPA's Office of Ground Water and Drinking Water's Standards and Risk Management Division: the Technical Support Center administers the UCMR program; the Targeting and Analysis Branch develops the CCL; and the Standards and Risk Reduction Branch leads the development of the regulatory determinations.

SDWA requires EPA to make regulatory determinations for at least five of the contaminants on the CCL every 5 years. Regulatory determinations can be positive (i.e., a primary standard is needed) or negative (i.e., no primary standard is needed). Under SDWA, EPA is to make a positive regulatory determination and subsequently issue a primary standard when (1) the contaminant may have an adverse effect on the health of persons; (2) the contaminant is known to occur or there is a substantial likelihood that the contaminant will occur in public water systems with a frequency and at levels of public health concern; and (3) in the sole judgment of the Administrator, regulation of such contaminant presents a meaningful opportunity for health risk reduction for persons served by public water systems. EPA can make a negative regulatory determination when the SDWA conditions are not met. SDWA set recurring 5-year deadlines for the CCL, UCMR, and regulatory determinations. The initial deadlines were staggered such that EPA was required to publish the CCL in the *Federal Register* first (February 1998), the UCMR list 18 months later (August 1999), and the regulatory determinations 2 years after that (August 2001). The cycle restarts with the next CCL, UCMR, and regulatory determinations due 5 years after the first set of deadlines. However, EPA has not always met the SDWA deadlines. Moreover, EPA has not issued the CCL, UCMR, and regulatory determinations separated by the 18-month and 2-year time frames that were originally established by statute.[15] The cycle is currently in its third iteration. EPA refers to the programs in the current cycle as CCL3, UCMR3, and the CCL3 regulatory determinations (see table 2).

The UCMR program's primary purpose is to inform regulatory determinations, but EPA can also use the UCMR data to develop health advisories for some of the contaminants it decides not to regulate. Health advisories provide information on contaminants that can harm human health and are known or anticipated to occur in drinking water. They provide technical guidance on health effects to assist EPA regional offices, states, and other public health officials. EPA has issued new or updated health advisories for nine contaminants it decided not to regulate.[16] The advisories identify concentrations of contaminants at which adverse health effects are not anticipated to occur over specific exposure durations (1 day, 10 days, several years, and a lifetime). Drinking water health advisories are not legally enforceable standards and, according to EPA, are subject to change as new information becomes available.

Table 2. EPA Programs to Address Unregulated Contaminants in Drinking Water

Program	Contaminant Candidate List (CCL)	Unregulated Contaminant Monitoring Rule (UCMR)	Regulatory determinations
Purpose	Identify unregulated contaminants that are known or anticipated to occur in public water systems	Monitor unregulated contaminants to determine the extent of their occurrence in drinking water	Decide whether or not to regulate contaminants on the CCL using UCMR data, health effects information, and other information
SDWA requirement	Publish list every 5 years	Publish a list of up to 30 contaminants for monitoring, every 5 years	Decide whether or not to regulate at least 5 CCL contaminants, every 5 years
Outputs	1998 (60 contaminants on CCL1) 2005 (51 contaminants on CCL2) 2009 (116 contaminants on CCL3)	1999 (26 contaminants on UCMR1) 2007 (25 contaminants on UCMR2) 2012 (30 contaminants on UCMR3)	2003 (9 negative determinations for CCL1) 2008 (11 negative determinations forCCL2) 2011 (1 positive determination–out of cycle) 2015 (determinations for CCL3 expected)

Source: GAO analysis of EPA information.

Laboratories and Methods Used to Analyze Drinking Water

Private analytical laboratories and some water utilities' laboratories conduct the analytical testing for the UCMR contaminants. All laboratories conducting analyses for UCMR3 must be approved by EPA. In general, laboratories must first successfully complete the registration and application process, which includes testing for contaminants using EPA-specified analytical methods, before receiving EPA's approval.[17] Approved laboratories must adhere to the quality assurance and quality control procedures and criteria outlined in the methods and rule language to maintain their approval status and may be subject to on-site laboratory or paper audits. Laboratories also post occurrence data and required quality control data electronically in EPA's data system, the Safe Drinking Water Accession and Review System, within 120 days of the sample collection date.[18]

EPA specifies the analytical methods that laboratories participating in the UCMR program use to detect contaminants. Some methods can detect a single contaminant while others can detect a group of similar contaminants. Because the UCMR collects nationwide data, agency officials told us EPA's goal is to use analytical methods that yield consistently high-quality monitoring data from public water systems across the country. EPA officials told us that the agency develops or uses analytical methods for the UCMR program that enable it to meet this goal. In particular, agency officials told us these methods must (1) have sufficient "holding times";[19] (2) be able to be used successfully by a variety of labs with varying levels of expertise; and (3) be used effectively in different types of drinking water. EPA officials said it typically takes 1 or 2 years to develop a standard analytical method that meets these requirements and up to 3 years or longer to develop more complex methods. See appendix III for more information on analytical methods.

EPA HAS IMPROVED THE UCMR PROGRAM BY IMPLEMENTING OUR RECOMMENDATIONS, BUT ITS MONITORING FREQUENCY MAY YIELD SOME INACCURATE ESTIMATES

EPA has implemented the three recommendations we made in our May 2011 report to improve the UCMR program.[20] The technical changes EPA has made to UCMR3 are significant improvements over prior UCMR cycles and have increased the program's productivity and statistical strength, among other things. However, the program's monitoring frequency may result in inaccurate estimates of the occurrence of some contaminants.

EPA Has Implemented All of GAO's 2011 Recommendations to Improve the UCMR Program

In UCMR3, EPA implemented our recommendations to (1) monitor for 30 contaminants; (2) monitor for most or all contaminants using the program's more robust monitoring approach; and (3) select minimum reporting levels (MRL) that are sufficiently sensitive to detect the presence of contaminants in public water systems at levels of public health concern as follows:[21]

Monitor for 30 Contaminants

In our May 2011 report,[22] we reported that EPA did not use its full statutory authority to monitor for 30 contaminants in implementing the first two cycles of the UCMR program. Specifically, EPA required that 26 and 25 contaminants be monitored under UCMR1 and UCMR2, respectively. As a result, EPA did not obtain occurrence data for 9 additional contaminants, which in turn, limited the agency's productivity in obtaining critical information to support its regulatory determinations. In its response to our recommendation that it use its full statutory authority, EPA stated it supports a goal of including as many priority contaminants as possible in each 5-year monitoring cycle. Accordingly, in May 2012, EPA published its final rule for UCMR3 requiring public water systems to monitor for 30 contaminants (see app. II for a list of the 30 contaminants).[23]

Use More Robust Monitoring Approach

In 2011, we reported that EPA officials described the assessment monitoring approach as "the gold standard" for obtaining sufficient data on national occurrence of drinking water contaminants to support regulatory determinations, but that it had used the more limited screening survey approach for a majority of the contaminants monitored under UCMR1 and UCMR2, as shown in table 3. The screening survey monitoring approach produces estimates of national occurrence with greater uncertainty than those provided by assessment monitoring because it requires a smaller sample of public water systems to monitor. For example, under UCMR1, the screening survey required a sample of 300 public water systems to monitor. In contrast, approximately 3,800 systems were required to conduct assessment monitoring. In EPA's response to our recommendation that it conduct monitoring for most or all contaminants using the assessment monitoring approach, EPA stated that it supports a goal of using assessment monitoring for as many contaminants as is practical to obtain more robust occurrence data for and also to achieve higher confidence in the national estimates. EPA is requiring assessment monitoring for a majority of the contaminants under UCMR3. In addition, for UCMR2 and UCMR3, EPA expanded the sample size of the screening survey to increase the statistical strength of the monitoring results. We analyzed EPA's methodology for selecting the sample of systems for the screening survey under UCMR2 and UCMR3 against standard statistical criteria, and we determined that the larger sample size should allow EPA to make national estimates of contaminant occurrence with smaller margins of error and higher levels of confidence compared with the first UCMR. In addition, 26 of the 40 experts who participated in our survey thought the screening survey monitoring approach for UCMR3 is likely to provide EPA with sufficient data to determine whether contaminants occur with a frequency and at levels of public health concern. Further, EPA officials told us that the UCMR2 screening survey occurrence data are sufficient for supporting regulatory determinations, whereas, in its 2008 regulatory determinations notice, EPA stated that the occurrence data were insufficient to support regulatory determinations for 12 of the 14 contaminants monitored under the UCMR1 screening survey.

Table 3. Number of Public Water Systems Required to Participate in the First, Second, and Third Unregulated Contaminant Monitoring Rules (UCMR1, UCMR2, and UCMR3)

Monitoring approach	UCMR1	UCMR2	UCMR3
Number of contaminants to be monitored			
Assessment monitoring	12	10	21[a]
Screening survey	14	15	7
Pre-screen testing	0	0	2
Total number of contaminants	26	25	30
Number of public water systems required to participate			
Assessment monitoring	~3,800[b]	~4,200[b]	~5,000[b]
Screening survey	300[c]	~1,200[c]	~1,200[c]
Pre-screen testing	NA	NA	800

Source: GAO analysis of EPA information.

[a]UCMR3 assessment monitoring includes total chromium, a regulated contaminant being monitored under another authority.

[b]Assessment monitoring includes all public water systems serving over 10,000 people and a representative sample of 800 systems serving 10,000 or fewer people.

[c]EPA increased the sample size of the screening survey for UCMR2 and UCMR3. Specifically, under UCMR1, the screening survey included a sample of 300 systems (120 systems serving over 10,000 people and 180 systems serving 10,000 or fewer people). In contrast, under UCMR2 and UCMR3, the screening survey includes all systems serving over 100,000 people; a representative sample of 320 systems serving 10,001 to 100,000 people; and a representative sample of 480 systems serving 10,000 or fewer people.

Select Sufficiently Sensitive MRLs

In 2011, we reported that the utility of the UCMR occurrence data was reduced because EPA had not used sufficiently sensitive MRLs to detect the presence of contaminants at levels of public health concern. Specifically, for 9 of the 20 contaminants for which EPA made regulatory determinations in 2003 and 2008, the MRL exceeded EPA's health reference level—the health benchmark EPA uses to assess whether the detection of a specific contaminant may indicate a potential human health risk.[24] As a result, occurrence of these 9 contaminants at a level higher than the health reference level, but lower than the MRL, may not have been detected. In its response to our recommendation that EPA use MRLs that are sufficiently sensitive, EPA stated it supports a goal of establishing minimum reporting levels that are sufficiently sensitive to reliably detect the known and likely occurrence of contaminants in public water systems at levels of public health concern and provide useful and credible information on the occurrence of the contaminants in public drinking water systems.[25] EPA has since revised its approach for selecting MRLs for monitoring selected contaminants in an effort to increase the likelihood that the monitoring will be sufficiently sensitive to detect the presence of contaminants at levels of public health concern. Specifically, EPA officials told us that, for UCMR3, the agency set the MRLs based on how low the method could reliably measure specific contaminants.[26] These levels are based on method sensitivity (i.e., not adjusted upward based on available health effects information). In some cases, the MRLs for UCMR3 are set well below EPA's current health reference levels, according to EPA's final rule for UCMR3. However, health effects research on some UCMR3 contaminants is still in progress, and the health reference levels are subject to change. EPA believes that its new approach will be more effective than its past approach because, by monitoring for contaminants at the lowest concentrations feasible, UCMR3 is more likely to detect contamination at levels of public health concern, even if new health effects data result in lower health reference levels in the future, as anticipated by its final rule for UCMR3.

UCMR Monitoring Frequency May Yield Inaccurate Estimates of Some Contaminant Occurrence

EPA has taken steps to improve the UCMR program, but the program's standard monitoring frequency may result in inaccurate estimates of the occurrence of certain contaminants. In particular, several experts we surveyed and studies on contaminant occurrence indicated that the UCMR3 standard monitoring frequency may result in inaccurate estimates of the occurrence of sporadically occurring microbes (e.g., viruses) and seasonally applied pesticides. SDWA calls for EPA to vary the monitoring frequency and schedule for public water systems based on three criteria (1) the number of persons served by the system, (2) the source of the system's water supply, and (3) the contaminants likely to be found. Under UCMR3, EPA varies the monitoring frequency for systems based on a system's size and water supply (e.g., surface water or groundwater), but it does not vary it based on the type of contaminant likely to be found. Specifically, under UCMR3, surface water systems and groundwater systems under the direct influence of surface water[27] are required to conduct quarterly monitoring of selected chemical contaminants, while groundwater systems are required to monitor selected chemical contaminants twice a year. In addition, nondisinfecting groundwater systems that serve 1,000 or fewer people and are located in sensitive hydrogeological areas are required to allow EPA monitoring for two microbial contaminants (viruses) twice a year (see table 4). The virus monitoring in UCMR3 was designed in consideration of where these viruses may be more likely to be found—small groundwater systems that do not disinfect. Nonetheless, EPA used the standard monitoring frequency for viruses; it did not vary the frequency or schedule based on the contaminants likely to be found. Similarly, contaminant monitoring under UCMR2 occurred quarterly for surface water systems and twice a year for groundwater systems for all contaminants. EPA did not vary the monitoring frequency or schedule during the seasons when pesticides were applied and the likelihood of occurrence was highest.

Table 4. Monitoring Frequency by System Size and Water Supply Source for the Third Unregulated Contaminant Monitoring Rule

System size	Water supply source	Monitoring frequency
Assessment monitoring		
All systems serving more than 10,000 people and a representative sample of 800 systems serving 10,000 or fewer people	Surface water or groundwater under the direct influence of surface water	Quarterly during a 1-year period with sampling events 3 months apart
	Groundwater	Twice in a 1-year period with samplingevents 5-7 months apart
Screening survey		
All systems serving more than 100,000 people, a representative sample of 320 systems serving 10,001 to 100,000 people, and a representative sample of 480 systems serving 10,000 or fewer people	Surface water or groundwater under the direct influence of surface water	Quarterly during a 1-year period with sampling events 3 months apart
	Groundwater	Twice in a 1-year period with samplingevents 5-7 months apart
Pre-screen testing		
A representative sample of 800 nondisinfecting groundwater systems serving 1,000 or fewer people	Groundwater	Twice in a 1-year period with samplingevents 5-7 months apart

Source: GAO analysis of EPA documents.

When EPA first established the UCMR program, it explained that the standard monitoring frequency for public water systems using surface water and groundwater under the direct influence of surface water will capture variability in contaminant occurrence because one of each public water system's quarterly sampling events must occur during what EPA considers to be the most vulnerable period for drinking water contamination, i.e., the period of peak contamination.[28] According to EPA's explanation in the *Federal Register* accompanying the final rule for UCMR1, studies show that the greatest risk of contaminant occurrence for much of the country is the late-spring/early-summer runoff period from May to July, particularly for contaminants such as pesticides and nitrate.[29] EPA designed the UCMR3 program to include sampling during this period, as designated by EPA or the state.

Quarterly monitoring for systems using surface water or groundwater under the direct influence of surface water occurs during one of three possible time frames (1) January, April, July, and October; (2) February, May, August, and November; or (3) March, June, September, and December. EPA assigns one of these three time frames to each participating public water system using surface water or groundwater under the direct influence of surface water.[30] EPA did not specify a vulnerable period for public water systems using groundwater in UCMR2 and UCMR3, but it stated that samples will be taken during all months and seasons of the year when all groundwater systems are considered together, according to agency documents.[31]

However, several experts we surveyed and studies on contaminant occurrence indicated that the UCMR standard monitoring frequency may not accurately characterize the presence of certain contaminants in public water systems. In particular, the monitoring frequency may result in inaccurate estimates of the occurrence of sporadically occurring microbes (e.g., viruses) and pesticides whose peak concentrations typically occur during and after the application season, primarily from April through July. For example, EPA's monitoring frequency may underestimate occurrence for compounds such as seasonally applied pesticides, according to one expert we surveyed.

As noted above, the UCMR3 standard monitoring frequency may result in inaccurate estimates of the occurrence of some microbes and pesticides.

- *Microbes*. Under UCMR3, EPA selected a representative sample of 800 nondisinfecting groundwater systems located in sensitive hydrogeological areas[32] to monitor for two viruses—enterovirus and norovirus—using the pre-screen testing monitoring approach.[33] In the notice for its proposed rule for UCMR3, EPA states that the agency limited its selection to those systems serving 1,000 or fewer persons because these smaller systems are the least likely to disinfect the water and, therefore, the most vulnerable to contamination with viruses. Monitoring and analyses are organized and paid for by EPA through direct contracts with laboratories. Under UCMR3, each of the selected systems will be monitored for the viruses twice during a 12- month period from January 2013 to December 2015. The systems' sampling events will be 5 to 7 months apart. However, more than half of the experts who completed our survey (23 of 40) said the twice-yearly monitoring requirement for viruses is not likely to provide EPA sufficient data to determine whether the contaminants occur with a frequency and at levels of public health concern. In addition, 15 experts who completed our survey stated that more frequent monitoring may be warranted to capture the occurrence of viruses in public water systems, which many characterized as variable. Similarly, a technical presentation that EPA used to support its rationale for including the viruses in UCMR3 shows that virus occurrence in groundwater systems is highly variable and that groundwater recharge events, like heavy precipitation, increase virus contamination levels.[34] The technical presentation concludes that more frequent sampling is necessary to characterize virus occurrence in public water systems. Moreover, in the past, EPA varied the monitoring frequency for a microbe that public comments on EPA's proposed rule for UCMR1 characterized as sporadically occurring.[35] Specifically, under UCMR1, surface water systems and groundwater systems monitoring for the microbial contaminant, *Aeromonas*, were required to take samples quarterly, with additional samples taken, so that samples generally were collected each month during the warmest months of the year (July, August, and September) because EPA officials told us that this is when they expected to find contaminant occurrence.[36] Systems monitored during one of three possible time frames (1) January, April, July, August, September, and October; (2) February, May, July, August, September, and November; or (3) March, June, July, August, September, and December. In contrast, UCMR3 does not require any additional monitoring for viruses. As a result, the data obtained under UCMR3 may not provide reliable estimates of the occurrence of viruses in drinking water.

- *Pesticides*. Under UCMR2, all public water systems serving 100,000 or more people and a representative sample of 800 systems serving fewer than 100,000 people were required to monitor for nine pesticides using the screening survey monitoring approach. These pesticides included three widely used herbicides applied for weed control on crops such as corn, cotton, and soybeans, and six degradates.[37] Contaminant monitoring occurred quarterly for surface water systems and biannually for groundwater systems. No additional monitoring was conducted during the seasons when pesticides were applied and the likelihood of occurrence was highest. Two of the 14 experts who participated in the first round of our survey said that the UCMR standard monitoring frequency may result in inaccurate estimates of the occurrence of such pesticides in public water systems. For example, one expert said that, following application, pesticides can occur at relatively high concentrations but for short periods of time, and that, because the UCMR's timing and frequency of monitoring are not based on this consideration, the program could completely miss that occurrence. Data from a United States Geological Survey (USGS) report also show that concentrations of herbicides are usually highest during and after the application season, primarily from April through July.[38] For example, at four stream sites studied, USGS found that the presence of atrazine, the dominant herbicide used in the Corn Belt during the study period, typically peaked after applications in the spring and gradually declined through the summer and winter. According to the USGS report, such seasonal patterns of pesticide concentrations are generally consistent each year, as long as the pesticides are still in use. Pesticide occurrence in groundwater also generally coincides with pesticide use patterns. Sampling was conducted throughout the year, including intensive sampling during the times of highest pesticide use and runoff. This sampling approach enabled USGS to conclude that concentrations of pesticides in streams and groundwater were typically below water-quality benchmarks for human health. Thus, seasonal monitoring of pesticides is important to assess whether pesticides occur at concentrations that may affect human health. Also, high-frequency sampling during seasons when intense pesticide use coincides with periods of high runoff may be needed to ensure that the highest pesticide concentrations are detected. In contrast, EPA's standard monitoring frequency requires most public water systems to monitor for contaminants only once per season or once over two seasons. As a result, some of the data collected under UCMR2 may not yield accurate estimates of pesticide occurrence in public water systems.

EPA CONSIDERED 10 FACTORS TO SELECT CONTAMINANTS FOR MONITORING, BUT ITS SELECTION PROCESS FACED LIMITATIONS

EPA used 10 selection factors, as well as a working group, and internal and external reviews, among other steps, to select contaminants for UCMR3. However, the process was lengthy due primarily to the reviews and the time required to develop analytical methods. The selection process was also limited by a statutory cap of 30 contaminants, which restricted EPA's ability to collect data on additional contaminants that could have been monitored for very little additional cost. The selection process also excluded contaminants previously monitored in earlier UCMR cycles, which has some limitations.

UCMR3 Contaminant Selection Considered Multiple Factors, but the Process Was Lengthy

To select contaminants for UCMR3, EPA considered 10 factors and used a prioritization process to identify potential contaminants, but took over 30 months, or 2 1/2 years, to develop and publish the final list in the *Federal Register* after the CCL was published. Had EPA consistently met the initial and recurring statutory deadlines for CCL and UCMR, they would have been separated by 18 months.

Contaminant Selection Factors

EPA used 10 factors to select contaminants for UCMR3 (see fig. 1). Officials told us that a contaminant does not have to meet all of the selection factors to be chosen; however, three are very important to consider as follows:

- *Analytical method availability.* Because analytical methods are the means by which laboratories test for contaminants, EPA officials told us they could not include a contaminant on the UCMR list without also having an analytical method available that is sensitive and precise enough to provide the data quality and accuracy that EPA needs for the UCMR program. See appendix III for more information on analytical methods.
- *Health effects information.* Because of the importance of monitoring for UCMR contaminants that may pose a public health risk and, therefore, may need to be regulated, EPA officials told us they placed a high priority on contaminants for which they have reliable health effects data.
- *Need for data to support regulatory determinations on priority contaminants.* Because the goal of the UCMR program is to provide occurrence data on unregulated contaminants for informing regulatory determinations, the Standards and Risk Reduction Branch's priorities for data to support a regulatory determination are heavily considered.

1. Whether an analytical method was available and sensitive enough to detect contaminants at the desired level;[a]
2. Whether information on the health effects of exposure to a contaminant was available and reliable;
3. Whether EPA's Standards and Risk Reduction Branch identified the contaminant as a priority for obtaining occurrence data in order to support a regulatory determination;
4. Whether the contaminant had already been monitored under a prior Unregulated Contaminant Monitoring Rule (UCMR); if so, it was eliminated from further consideration;[a]
5. Whether the contaminant is listed on the Contaminant Candidate List, which serves as the primary basis for developing the UCMR list;[a]
6. Whether EPA identified it as a contaminant of emerging concern;[a]
7. Whether industry trends show an increase or decrease in the use of certain chemicals that could be contaminants;
8. Whether stakeholders' input through EPA's public meetings, UCMR3 working group, and public comments on the proposed UCMR rule identify a contaminant as a priority;
9. Whether a contaminant could be monitored using the same analytical method being used for contaminants already chosen for the UCMR, which would make monitoring it cost-effective; and
10. Whether monitoring data from other organizations, such as USGS and states, indicate that the contaminant is likely to be present in drinking water.

Source: GAO analysis of EPA information.

[a] EPA identified these selection factors in the *Federal Register*. EPA officials identified the other factors during interviews with us.

Figure 1. EPA Selection Factors for Identifying Contaminants for Monitoring under the Third Unregulated Contaminant Monitoring Rule (UCMR3).

Contaminant Selection Process

According to EPA documents and officials, EPA used a prioritization process to select potential UCMR3 contaminants. First, EPA considered the contaminants listed on CCL3 and identified contaminants of emerging concern[39] to establish an initial list of about 150 contaminants. Contaminants for which analytical methods were not available were then eliminated from consideration for UCMR3, however EPA officials told us that those contaminants may be monitored under a future UCMR if an analytical method can be developed. Contaminants that were monitored under UCMR1 or UCMR2 were also eliminated from further consideration early in the UCMR3 selection process, as reported in the *Federal Register* and agency records.[40] These two selection factors reduced the list to fewer than 35 contaminants, according to EPA documentation. In addition, officials told us they met regularly with representatives from the Standards and Risk Reduction Branch to ensure the list of UCMR contaminants would meet their needs for data to support regulatory determinations.

EPA finalized the UCMR3 list through deliberations with other EPA staff and stakeholders and by holding internal and external reviews, among other steps (see fig. 2). In November 2009, a UCMR3 working group led by EPA's Technical Support Center and composed of EPA national and regional staff, as well as representatives from states, began to select the final 30 contaminants. The EPA Office of Science and Technology[41] conducted health effects evaluations for possible UCMR3 contaminants, according to EPA documents and agency officials. The UCMR3 working group, whose members included experts from the Office of Science and Technology, reviewed this health effects information. For each UCMR3 chemical contaminant, EPA developed an information sheet that listed known health effects information and the sources of that information, among other things. EPA used these contaminant information sheets for internal considerations and published them in the EPA docket for UCMR3. After the working group discussed the list of contaminants, the Technical

Support Center submitted the proposed and final UCMR contaminant lists and associated rules to EPA management for their review. The EPA internal reviews involved different offices within EPA, including the Office of the EPA Administrator. EPA also submitted the proposed and final UCMR lists and rules to the Office of Management and Budget (OMB) for its review. On March 3, 2011, EPA published a proposed list of 30 contaminants and the associated rule in the Federal Register. EPA published the final UCMR3 contaminant list and associated rule on May 2, 2012 (see app. II for list of UCMR3 contaminants).

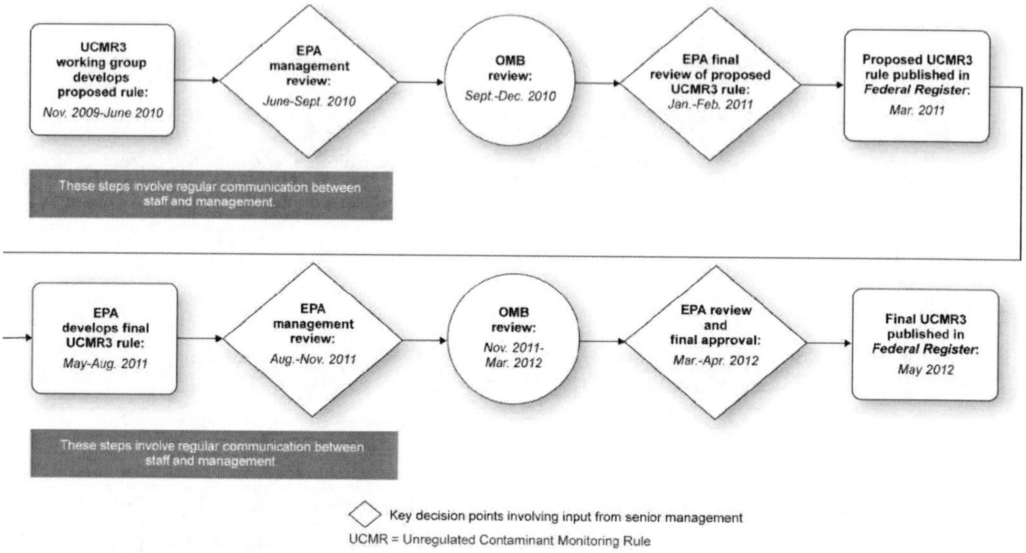

Source: GAO analysis of EPA information.

Figure 2. Contaminant Selection and Rule Development Process for the Third Unregulated Contaminant Monitoring Rule.

The UCMR3 selection process led EPA to select many contaminants from the CCL, EPA's list of the unregulated drinking water contaminants that may require regulation because they are known or anticipated to occur in public water systems. Specifically, 22 of the UCMR3 contaminants were also listed on CCL3. The CCL3 process was developed using substantial expert input and recommendations from various groups, including the National Academy of Science's National Research Council, the National Drinking Water Advisory Council, and EPA's Science Advisory Board, according to EPA documents. In addition, 37 of the 40 experts who participated in our survey said that CCL3 was a reasonable source for EPA to use when it began identifying contaminants for UCMR3. UCMR3 also included 7 contaminants of emerging concern that were not on CCL3, and total chromium.[42] Further, 33 of the 40 experts said it was worthwhile for EPA to select contaminants for the UCMR that are not on the CCL. For example, one expert in our survey told us that including such contaminants is the best way to identify emerging contaminants that come to light between the publication of the CCL and UCMR list. Overall, 30 of the 40 experts who participated in our survey reported that the contaminants EPA selected for UCMR3 warrant monitoring.

Contaminant Selection Process Was Lengthy

EPA's UCMR3 contaminant selection and rule development process took more than 2 1/2 years to complete. EPA started the formal UCMR3 selection process in November 2009 but did not publish the final UCMR3 list until May 2012. SDWA specifies that EPA publish a new UCMR list every 5 years. Whereas EPA's previous list was issued in 2007, it met the 5-year time frame required by SDWA. However, the UCMR program is no longer staggered 18 months after the CCL, as SDWA intended. Rather, UCMR3 was issued 31 months after CCL3. Based on the length of its process, EPA would have had to begin selecting contaminants in September 2008 to have issued UCMR3 within 18 months of CCL3. Further, EPA cannot implement the UCMR program until it publishes the final UCMR list and rule in the *Federal Register*, which delays the onset of the data collection period. This delay, in turn, delays the availability of the UCMR data for making regulatory determinations (see fig. 3).

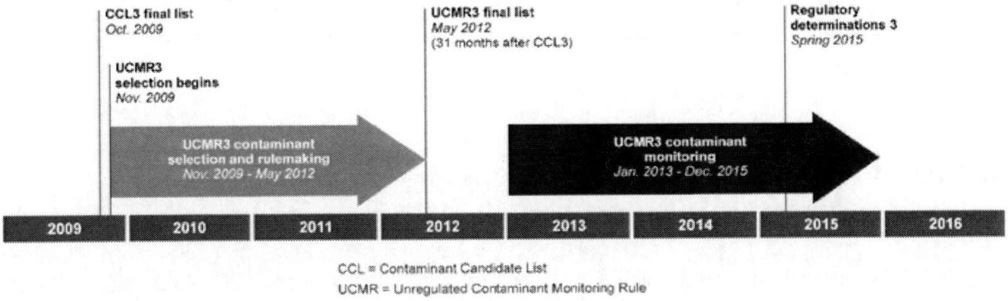

Source: GAO.

Figure 3. UCMR3 Timeline.

The UCMR3 contaminant selection and rulemaking process concluded 31 months after the CCL was issued primarily because EPA waited to initiate its working group until CCL3 was finalized, and the EPA and OMB reviews were lengthy. First, EPA's UCMR3 working group did not begin its work until November 2009, after the final CCL3 had been published in October 2009. EPA officials told us that discussions about potential contaminants for UCMR3 and analytical methods that were likely to be available to detect them began earlier, but they waited until November 2009 to begin the formal selection process, in part, because of staffing needs. The staff who developed UCMR3 were also coordinating and providing technical support to the public water systems that were monitoring for contaminants under UCMR2, which took place between January 2008 and December 2010. In addition, CCL staff serve on the UCMR working group, and UCMR staff also serve on the CCL working group, which EPA officials told us limits their availability. Second, EPA internal reviews, OMB reviews, and the revisions to the UCMR3 proposed and final rules prompted by the reviewers' comments added 16 months in total to the process. However, these reviews are not always required for this type of activity, according to agency guidance. Specifically, EPA's UCMR is conducted under the agency's Action Development Process, a process used for developing rules. The agency incorporated EPA management and OMB reviews into their process; however, these reviews are not always required for a program like the UCMR, according to agency guidance, as stated earlier.[43]

In addition, the EPA analytical method development process is complex, technically challenging and lengthy, which agency officials also told us presents a challenge for EPA to meet its deadline. EPA officials said it typically takes 1 to 2 years to develop a standard method that meets the requirements needed for the UCMR program and up to 3 years or longer for more complex methods. In addition, EPA generally has not used a specific analytical method in more than one UCMR cycle, although some of those older methods could be used to monitor unregulated contaminants for which EPA does not yet have UCMR occurrence data. For example, an analytical method used under UCMR1 could have detected seven CCL3 contaminants, but this method was not included in UCMR3. According to industry organizations and EPA officials, EPA did not formally reach out to stakeholders when developing analytical methods for UCMR 1, 2, or 3. However, EPA is formally reaching out to stakeholders to get their input on the development of analytical methods that may be used for UCMR4. In May 2013, EPA held a public meeting and webinar to provide stakeholders with an update on analytical methods that EPA has completed or has under development for potential use in the UCMR program and requested input from attendees to help address some of the technical challenges it faces.

Statutory Cap of 30 Contaminants Limits EPA's Ability to Obtain Additional Data at Little Extra Cost

The statutory cap limiting the UCMR program to 30 unregulated contaminants every 5 years (i.e., each UCMR cycle) impedes EPA's ability to obtain occurrence data on additional contaminants that could be monitored for little additional effort or cost. Many of the analytical methods that EPA is using for UCMR3 are able to test a single sample of drinking water for more than one contaminant at a time. Therefore, data on additional contaminants could be obtained without additional sampling by the public water systems. EPA is allowing laboratories to use eight EPA-developed analytical methods to test for UCMR3 contaminants and four methods developed by others. These methods can detect a total of 67 unregulated contaminants, including 4 listed on CCL3 and 8 contaminants of emerging concern, according to our analysis of EPA data. There isbroad recognition throughout the federal government that agencies need to operate programs efficiently and maximize efficiencies when possible.[44] In addition, EPA officials who use the UCMR data to make regulatory determinations, while sensitive to the practical limitations of monitoring for additional contaminants, said that it would be helpful to have additional UCMR data if a method already in use for UCMR could detect the occurrence of additional contaminants. However, because of the SDWA limit of 30 contaminants, EPA cannot always take advantage of this efficiency and is unable to gain economies of scale by monitoring for additional contaminants using monitoring that is already under way.

Specifically, our analysis indicates that four CCL3 chemicals—1,1,1,2-Tetrachloroethane; n-propylbenzene; sec-butylbenzene; and methyl tertiary-butyl ether (MTBE)—could have been detected using one of the UCMR3 analytical methods. As noted earlier, CCL3 represents EPA's list of the unregulated drinking water contaminants that may require regulation because they are known or anticipated to occur in public water systems. EPA had included n-propylbenzene and sec-butylbenzene in the proposed list of UCMR3 contaminants, but it did not include them on the final UCMR3. Instead, EPA included chromium-6, as well as total chromium, which EPA stated became higher priorities for monitoring for the agency. Public comments on the draft UCMR3 list also supported monitoring these two contaminants. EPA's Standards and Risk Reduction Branch, which develops the regulatory determinations, supported the inclusion of the two chromiums, as did a majority of the experts who participated in our survey (35 of 40 supported adding chromium-6 and 28 of 40 supported adding total chromium). However, by including these two higher priority contaminants, EPA removed two others in light of the cap of 30 contaminants and could not take advantage of economies of scale by using a method already selected for UCMR3 to obtain additional data. In addition, MTBE could also have been monitored using the same method. However, EPA officials told us they decided not to include MTBE, which was monitored under UCMR1, because they did not need additional information about it. As noted earlier, the priorities of EPA's Standards and Risk Reduction Branch are a key selection factor for which contaminants to include on the UCMR. In the absence of the limit of 30, EPA would have had an opportunity to monitor for these CCL3 chemicals, as well as 1,1,1,2-Tetrachloroethane, as part of UCMR3 at nominal extra cost. EPA officials told us that adding a contaminant to an ongoing analysis typically costs about $5 to $10.

Similarly, our analysis indicates that eight contaminants of emerging concern known as perfluorinated chemicals, or PFCs, could also have been detected by the same analytical method that EPA is using to monitor six other PFCs (see table 5).[45] In 2009, EPA published an action plan to address some of these chemicals, which the agency describes as persistent in the environment and in people.[46] This action plan also states that it can reasonably be anticipated that continued exposure to PFCs could harm human health. Due to industry's increasing use of various PFCs, EPA included six of the chemicals identified in this plan in UCMR3, according to EPA officials. However, according to our analysis, five additional PFCs from the plan are among the eight that could have been detected using the same analytical method but are not on UCMR3. EPA selected the six PFCs it thought most important, but the statutory limit of 30 contaminants curtailed EPA's opportunity to gather occurrence data on eight additional contaminants of emerging concern at little additional cost or effort. EPA could have included the additional PFCs in UCMR3 but doing so would have reduced the number of contaminants selected from CCL3, EPA's list of the unregulated drinking water contaminants that may require regulation because they are known or anticipated to occur in public water systems. Nonetheless, the agency will not have UCMR data on whether or not those eight contaminants are present in the nation's drinking water until the next UCMR cycle at the very earliest; UCMR4 data collection is expected to be completed in 2021. However, EPA has not decided which contaminants will be monitored under UCMR4, and so it remains to be seen if they are included.

The legislative history of the SDWA Amendments of 1996 generally noted concerns with the financial limitations of public water systems and their abilities to absorb costs for activities such as monitoring. While the number of contaminants being monitored is a factor, costs related to analytical methods, such as laboratory equipment and labor, are key cost drivers for UCMR monitoring, according to EPA officials. EPA officials told us that they carefully consider the cost on public water systems when selecting analytical methods for the UCMR program. In addition, EPA pays for some of the small public water systems' monitoring and must stay within its budget for those costs.[47] EPA documentation indicates that UCMR3 will generally cost public water systems $50 to $470 per water sample, per analytical method. EPA officials told us that the cost to detect additional contaminants for an analytical method already in use was nominal—typically about $5 to $10 per water sample, per contaminant. Thus, the cost of adding contaminants to an analytical method that is already being used would likely pose only a modest cost increase for public water systems.

Table 5. Contaminants of Emerging Concern Detectable with an Analytical Method Already in Use

Contaminants of emerging concern
N-Ethyl perfluorooctanesulfonamido acetic acid
N-Methyl perfluorooctanesulfonamido acetic acid
Perfluorodecanoic acid
Perfluorododecanoic acid
Perfluorohexanoic acid
Perfluorotetradecanoic acid
Perfluorotridecanoic acid
Perfluoroundecanoic acid

Source: GAO analysis of EPA data.

Note: All of the contaminants of emerging concern listed in this table are perfluorinated chemicals, or PFCs. PFCs are man-made chemicals that are used in a wide variety of industrial and commercial products. EPA has reported that PFCs have significant adverse effects on laboratory animals and wildlife and anticipates that continued exposure to PFCs could result in adverse effects on humans.

UCMR3 Excludes Previously Monitored Contaminants

One of EPA's key selection factors for UCMR3 involved generally eliminating contaminants that were monitored under a prior UCMR round,[48] but by eliminating those contaminants, EPA may encounter data limitations because (1) the sampling methodology was not sufficient, (2) the health information has been updated, or (3) the data have become outdated.

Sampling Methodology

One of the sampling methodologies EPA used in a prior UCMR did not provide sufficient occurrence data to make regulatory determinations for some contaminants, according to EPA's regulatory determinations for the CCL2 contaminants.[49] Specifically, UCMR1 used a limited sample size of 300 public water systems for the 14 contaminants it monitored using the screening survey approach. This resulted in occurrence data for some contaminants that, according to EPA documents, were not sufficient to make a national estimate. In its July 2008 Regulatory Determination notice, EPA found that a total of 12 contaminants monitored under UCMR1 had insufficient occurrence data for making national estimates. Three of those contaminants are also listed on CCL3—the pesticides disulfoton, diuron, and terbufos. By applying the UCMR selection factor to eliminate contaminants monitored under previous UCMR rounds, EPA did not consider remonitoring these contaminants in UCMR3, even though EPA previously found the existing data were not sufficient to make national estimates of contaminant occurrence. However, 31 of the 40 experts who participated in our survey said it would be worthwhile to monitor UCMR contaminants again if they had insufficient data due to the sampling size.

According to EPA's notice in the *Federal Register* for UCMR1, EPA's original intent was to use the screening survey as a first indicator of potential contamination and then to remonitor the contaminants again using assessment monitoring to obtain nationally representative results if the contaminants occurred frequently enough in drinking water to warrant doing so.[50] However, EPA officials told us they later determined that this approach was not feasible because of the time it required and did not follow through with its plan to remonitor the UCMR1 contaminants. For the UCMR2 and UCMR3 cycles, EPA revised its approach and increased the screening survey sample size from 300 to 800. Specifically, under UCMR1, the screening survey included a sample of 300 public water systems.[51] Under UCMR2 and UCMR3, the screening survey included a sample of 800 public water systems and all of the systems serving more than 100,000 people, for a total of approximately 1,200 systems.[52] This increase allowed EPA to make nationally representative estimates based on the occurrence data.

Health Effects Information

According to EPA officials, health effects information for unregulated contaminants is often uncertain and in flux because research is often being conducted at the same time UCMR monitoring is occurring. Those officials said the new research sometimes indicates that health effects occur at lower levels of exposure than previously thought. In recognition of this, EPA set the monitoring levels, or MRLs, for UCMR3 based on how low the analytical method could reliably measure specific contaminants, as reported earlier. However, EPA previously set the MRLs for UCMR1 and UCMR2 with consideration of the levels of public health concern that were known at the time. In May 2011, we reported that EPA set the MRLs for some of the UCMR1 and UCMR2 contaminants before it had developed the health reference levels for those contaminants. EPA subsequently determined that those MRLs exceeded the health reference levels. As a result, occurrence of those contaminants at a level higher than the health reference level but lower than the MRL might not have been detected, and UCMR1 and UCMR2 occurrence data might not accurately capture the risk these contaminants pose to public health. Similarly, if the health effects information about the UCMR1 and UCMR2 contaminants changes in the future, EPA may not have the data it needs to support regulatory determinations. Thirty-two of the 40 experts who participated in our survey said it would be worthwhile to monitor again for UCMR1 and UCMR2 contaminants if new health effects information becomes available.

Age of Data

The UCMR data estimated contaminant occurrence at the time that they were collected. Due to changing industrial and agricultural chemical use and environmental conditions, UCMR data may not present an accurate picture of current contaminant occurrence after several years. In particular, the UCMR1 data, which were collected from 2001 to 2003, are now over 10 years old. Nineteen of the 40 experts who participated in our survey said that UCMR data this old are not likely to be a reliable indicator of the frequency and level of drinking water contamination, and 18 said they did not have enough information to judge how long the UCMR data are likely to be a reliable indicator. EPA officials and experts told us the shelf life of the UCMR data generally depends on the type of contaminant and whether its use is likely to have changed over time. According to the officials, concentrations of naturally occurring contaminants remain relatively constant in drinking water over time but concentrations of pesticides and industrial chemicals are subject to change. While some of the UCMR1 data may still be a reliable indicator of contaminant occurrence, several of the UCMR1 contaminants are pesticides and industrial chemicals and the extent of their use has changed since the UCMR data were collected. For example, use of the pesticides disulfoton and molinate effectively ended following EPA actions in 2009 and 2004, respectively.[53] Use of the pesticides terbufos and diuron also decreased significantly from 2003 to 2009, according to USGS data. In addition, use of two industrial chemicals, MTBE and nitrobenzene, decreased over time, according to EPA data. MTBE, a former gasoline additive, has not been used in significant quantities since 2005. Nitrobenzene use has fluctuated over time but decreased sharply from 1998 to 2005.[54] Because the UCMR1 data estimate contamination at a point in time prior to these changes, they likely overestimate current levels of contaminant occurrence.

EPA might choose not to make regulatory determinations on some contaminants if it concludes that the available occurrence data do not provide sufficient support, as it did for several contaminants in 2008. As we reported in May 2011, EPA did not make regulatory determinations for some contaminants monitored under a prior UCMR round because the occurrence data were not sufficient to do so.[55] SDWA requires EPA to use the best available data,[56] but by not remonitoring contaminants when appropriate, such as when the data collected in an earlier cycle have become outdated, EPA may not have sufficient data to support its regulatory determinations. SDWA does not prohibit EPA from remonitoring UCMR contaminants; therefore, the agency has the option to do so, for example when it believes that remonitoring would provide better support for its regulatory determinations. EPA officials told us that when developing future UCMRs, they will consider whether adequate occurrence data for a contaminant had already been collected under a prior UCMR, rather than generally eliminating all previously monitored contaminants as part of its contaminants selection process. They also said that if the officials who make the regulatory determinations asked for a contaminant to be remonitored, they would consider doing so.

UCMR Data Support Regulatory Determinations, but Time Lag Results in Delays and Reliance on Older Data

EPA uses the UCMR data to support regulatory determinations but faces a time lag when doing so. Agency officials told us UCMR data are EPA's primary source of data on the occurrence of unregulated contaminants in drinking water and are a key input for regulatory determinations. However, by the time the complete set of UCMR data is available, the recurring 5-year deadline SDWA established for making at least 5 regulatory determinations has passed, leading EPA, in some cases, to rely on older data that may no longer accurately reflect current contaminant occurrence in drinking water.[57]

UCMR Data Are EPA's Primary Occurrence Data Source and a Key Input for Regulatory Determinations

EPA uses the UCMR data, along with information about the contaminants' health effects and other information, to make regulatory determinations—that is, decisions on whether or not new primary drinking water standards are needed for particular unregulated contaminants. Regulatory determinations can be positive (i.e., a new primary standard is needed) or negative (i.e., no new primary standard is needed).[58] While SDWA requires EPA to make regulatory determinations for at least five CCL contaminants every 5 years, EPA officials told us their goal is to make as many regulatory determinations as possible, given the available data.[59] EPA used UCMR1 data to support 9 of the 11 regulatory determinations it made in 2008 for the CCL2 contaminants and another regulatory determination in 2011. It is currently using UCMR1 and UCMR2 data to inform its regulatory determinations for the CCL3 contaminants, which EPA expects to issue in 2015.

The complete UCMR3 data set will not be available until after those regulatory determinations have been issued. However, if the timing allows, EPA officials told us they may have an opportunity to use a partial UCMR3 data set to corroborate other data they have on contaminant occurrence when making the final regulatory determinations.

EPA officials told us that, when they are available, UCMR data are EPA's primary data source on unregulated contaminants in drinking water because they allow EPA to make national estimates of drinking water contamination, as discussed earlier in this report. EPA also has data on unregulated contaminants from other sources, such as USGS and some states, but those data do not provide nationally representative estimates of drinking water contamination. For example, the USGS data estimate contamination in source water (i.e., untreated water from rivers, lakes, and groundwater aquifers). Since source water can become public drinking water, the USGS data can indicate potential drinking water contamination. However, these data do not provide conclusive information on whether contamination is occurring in public water systems. EPA officials told us they use the other data to provide additional support for regulatory determinations, but they generally do not provide sufficient support on their own.

In contrast, the UCMR data are generally sufficient for estimating nationwide occurrence of drinking water contamination, with the exceptions noted earlier in this report. In addition, 35 of the 40 experts who participated in our survey said that the UCMR program has been an effective way to collect sufficient data to determine whether contamination is occurring frequently and at levels of public health concern in drinking water. Several experts provided additional comments on the overall effectiveness of the UCMR program. For example, one expert said that considering the great need for data and EPA's limited resources, the UCMR program has been one of the more effective tools in the agency's rule development process. Another expert said that, given statutory requirements, resources, and data availability, EPA has done an admirable job with the UCMR program.

Finally, one expert said that the UCMR program has been an effective way to remove contaminants from the radar screen when they did not warrant regulation.

Time Lag Delays Regulatory Determinations by a Full 5-Year Cycle and Results in the Use of Older Data

A time lag between the SDWA deadline for regulatory determinations and the UCMR data's availability typically delays determinations on given contaminants by a full 5-year cycle. SDWA sets deadlines for publishing the CCL and UCMR contaminant list and making regulatory determinations, but it does not set a deadline for monitoring the UCMR contaminants. SDWA established recurring 5-year deadlines for making regulatory determinations and for publishing the UCMR list. EPA has established a 3-year monitoring period for the UCMR, but this overlaps with the time frame during which EPA makes its regulatory determinations. EPA also uses some older data that may no longer accurately reflect current contaminant occurrence in drinking water when it is making regulatory determinations.[60]

Time Lag Typically Delays Regulatory Determinations for Contaminants by a Full 5-Year Cycle

A time lag between the statutory deadline for regulatory determinations and the UCMR data's availability typically delays regulatory determinations on given contaminants until the following cycle. SDWA directs EPA to publish a new CCL, UCMR, and set of regulatory determinations every 5 years, as described earlier in this report. The initial timeline staggered the deadlines such that EPA was required to publish the UCMR list 18 months after the CCL and issue the regulatory determinations 2 years after that. However, in practice, the 2-year window SDWA set from the time EPA publishes the UCMR list to when it makes regulatory determinations has not provided enough time for the agency to monitor the contaminants and incorporate the UCMR data into the determinations. Consequently, UCMR data are not available for EPA to make regulatory determinations for contaminants during the cycle in which they are monitored; rather, UCMR data for contaminants in one cycle typically are not used until the next cycle or later (see fig. 4).

EPA officials told us that the timing of the UCMR program is a key challenge, particularly in relation to the timing of the CCL and regulatory determination deadlines. Since the UCMR program began, the full UCMR data set has not been available until a year or more after EPA made its regulatory determinations for the same 5-year cycle. The time lag is influenced by the recurring 5-year deadlines for the CCL, UCMR, and regulatory determinations and increased when EPA missed the deadlines for CCL2 and UCMR2. As a result, UCMR2 data were not available until nearly 3 years after EPA made the regulatory determinations for the CCL2 contaminants. When EPA missed the UCMR2 deadline, subsequent deadlines were moved to a later date, consistent with the 5-year cycle established by SDWA.[61] Accordingly, the UCMR3 data will lag behind the regulatory determinations for the CCL3 contaminants by over a year. Based on EPA's current schedules for the three programs, the UCMR4 data will likewise lag behind the regulatory determinations for the CCL4 contaminants. EPA's schedule indicates that the agency plans to make regulatory determinations for the CCL4 contaminants in spring 2020 and complete UCMR4 monitoring in December 2020, with the full UCMR4 data set available in 2021. Without congressional action to make the time frames more realistic, the time lag and resultant delays are expected to continue in subsequent cycles.

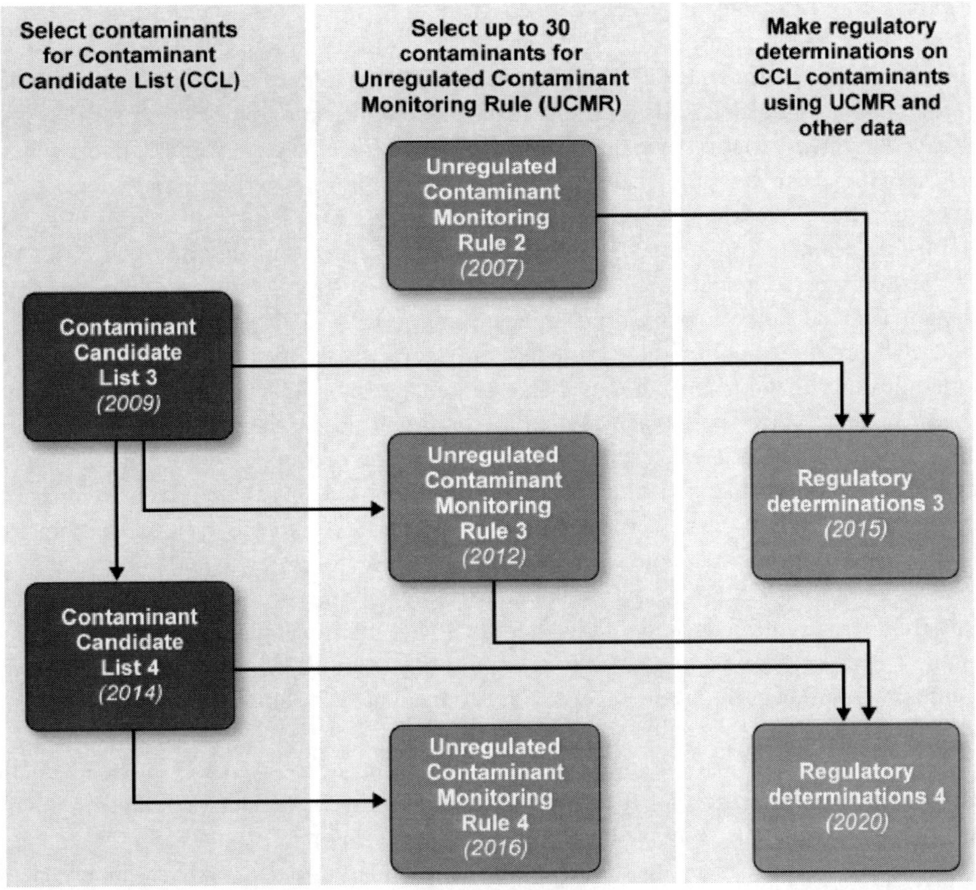

Source: GAO analysis of EPA data.

Figure 4. UCMR Data Do Not Inform Regulatory Determinations in the Same Cycle.

Three-Year Monitoring Period Contributes to Time Lag

EPA has established a 3-year monitoring period for the UCMR, which contributes to the time lag between the SDWA deadline for regulatory determinations and the UCMR data's availability. Before the monitoring period begins, EPA spends about 1 year preparing for the monitoring that public water systems will conduct. During this time, EPA finalizes the monitoring schedules for systems that will be participating in the program and approves the laboratories that will conduct the analyses, among other things. Each public water system monitors for contaminants during a 12- month window throughout the 3-year period that follows. EPA officials told us they spend up to 1 year finalizing the data collection and analysis, which includes about 6 months for public water systems and laboratories to report, review, and approve their data.[62] The UCMR3 monitoring, data collection, and analysis overlap with the time frame during which EPA will be making its regulatory determinations for contaminants on CCL3. UCMR3 monitoring is occurring from January 2013 through December 2015. Regulatory determinations for the CCL3 contaminants are expected in spring 2015, according to agency officials.[63] Officials who make the regulatory determinations told us that it is helpful to have the UCMR data about 2 years before publishing the final regulatory determinations. They also told us they start using the data once about 70 percent of the data set is available, approximately 2 years into the 3-year monitoring period. For example, EPA began using the UCMR1 data set to inform its internal deliberations on the regulatory determinations for CCL2 after about 70 percent of the data were available, according to agency officials. The officials said that by this time in the data collection period they saw very little variation in contaminant occurrence (i.e., the data were relatively stable).

EPA officials told us they selected the 3-year monitoring period mainly for logistical reasons impacting the agency and the public water systems. They told us that the 3-year monitoring period spreads the agency's work out over a more manageable period of time. EPA staff provide technical support to the public water systems conducting the monitoring, and many also have other responsibilities, according to EPA officials. The 3-year monitoring period also provides public water systems with more flexibility to decide when they will start monitoring. EPA officials told us that, after the agency publishes the final UCMR in the *Federal Register*, they assign each system a month and year to start monitoring. However, some systems request a later start date to incorporate the monitoring into their budgets and operations, according to agency officials. EPA officials also said laboratory capacity is sufficient under the 3-year period and has approved 53 laboratories to participate in UCMR3. EPA officials had told us that about 15 laboratories could probably handle the analytical work for UCMR3, therefore, laboratory capacity may be sufficient under a shorter monitoring period. Officials also told us that the 3-year monitoring period is not a given. For example, officials told us that a 2-year monitoring period, while challenging, may be possible for future cycles, but they would have to evaluate the shorter time frame in detail to determine its feasibility.

UCMR program officials also told us that monitoring over 3 years is intended to allow them to account for the effects that extreme weather in a given season could have on the data. They told us they did not know whether a shorter monitoring period, such as a 2-year period, would also allow them to do this because the extent of seasonal variability in the UCMR program is unknown. However, they did not expect that a shorter monitoring period would address variability to the same extent that the current 3-year monitoring period would because the longer period of time provides greater opportunity to detect it. Nonetheless, as reported earlier, EPA officials who make the regulatory determinations observed very few changes in occurrence trends after about 70 percent of the UCMR data had been collected, about 2 years into the 3-year monitoring period. When asked, experts who participated in our survey had mixed views on whether a 2-year monitoring period would be likely to provide EPA with sufficient information on seasonal fluctuations.[64] Specifically, 19 of 40 respondents thought that it was likely to, 11 thought it was not likely to, and 9 said they did not have enough information to judge. The UCMR3 data set may provide EPA with information that it can use to assess the extent of seasonal variability, according to agency officials. Because UCMR3 is monitoring for contaminants at low MRLs, it will likely have a higher rate of detections than prior UCMRs and, therefore, more data points that EPA can use to detect seasonal variability that it would have missed in the past. EPA could use this information to help it understand the extent of seasonal variability and the feasibility of a shorter monitoring period.

EPA Uses Occurrence Data from Previous UCMR Cycles When Current Data Are Not Available

EPA has adapted to the time lag by using UCMR data from previous cycles to make regulatory determinations. For example, EPA is using UCMR1 and UCMR2 data to make regulatory determinations for the CCL3 contaminants.[65] Similarly, EPA used UCMR1 data to make regulatory determinations for the CCL2 contaminants. Because EPA has carried over many contaminants from one CCL to the next, officials told us this is a workable approach.[66] However, more than half of the experts (25 of 40) who participated in our survey did not think using data from a prior UCMR cycle was an effective strategy for making regulatory determinations. Many of those experts were concerned because the time lag delays regulatory determinations on given contaminants and leads EPA to use older data that may not reflect current contaminant occurrence in some cases. The time lag delays regulatory determinations on contaminants by at least one 5-year cycle, and sometimes longer. EPA officials told us that most of the UCMR3 data will be used to support regulatory determinations for the CCL4 contaminants, which are expected to be issued in 2020. Those UCMR3 contaminants that are carried over to CCL4 will wait about 10 years for regulatory determinations unless EPA uses data from another source or makes an out-of-cycle regulatory determination (see fig. 5).[67]

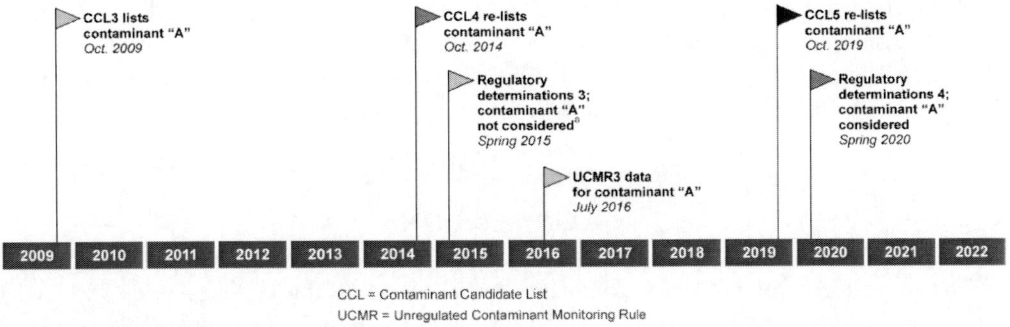

Source: GAO analysis of EPA data.
[a]Contaminant "A" not considered because UCMR data are not yet available.

Figure 5. About 10 Years Typically Pass before Regulatory Determinations Are Made for Contaminants on the CCL.

Regulatory determinations on given contaminants can also take longer to make. For example, 11 of the contaminants EPA included on a short list of potential regulatory determinations for CCL3[68] were first listed on CCL1 in March 1998.[69] By the time EPA makes its regulatory determinations, about 17 years will have passed since they were identified as contaminants known or anticipated to occur in public water systems. In addition, there are six contaminants that EPA carried over from CCL1 to CCL3 that it has not yet included in a regulatory determinations cycle. UCMR data are not yet available for any of them, although two—methyl bromide and 1,1-dichloroethane—are currently being monitored under UCMR3.[70] Methyl bromide is a pesticide that was phased out of use in 2005; 1,1,-dichloroethane is an industrial chemical. Due to the time lag, EPA's first use of the UCMR3 data for these CCL1 contaminants will likely be when it makes the regulatory determinations for CCL4.[71] This is about 22 years after they were first identified on the CCL as unregulated contaminants that are known or anticipated to occur in public water systems and may require regulation.

EPA is also using some older UCMR data that may not accurately reflect current contaminant occurrence in drinking water to make regulatory determinations, according to agency documents. Specifically, 6 of the 32 contaminants EPA included on its short list of potential regulatory determinations for CCL3 were monitored from 2001 to 2003 under UCMR1.[72] EPA officials told us that one of the reasons the UCMR1 data were not used for the regulatory determinations it issued in 2008 was because health effects information on some of the contaminants was not available.[73] EPA's July 2008 regulatory determinations indicate that there were health effects information gaps for 3 of the 6 contaminants.[74] As noted earlier in the report, the UCMR1 data are now over 10 years old and may not accurately reflect current levels of drinking water contamination in some cases. In particular, the 6 UCMR1 contaminants on EPA's short list are industrial chemicals and pesticides, and the extent of their use has changed since the UCMR data were collected, as described earlier. Four of the 6 contaminants are pesticides: disulfoton, molinate, terbufos, and diuron. The other 2 contaminants are industrial chemicals: MTBE and nitrobenzene. Further, in its July 2008 Regulatory Determination notice, EPA stated that it had insufficient occurrence data for making national estimates of disulfoton, diuron, and terbufos contamination. Occurrence data had been collected on those contaminants under UCMR1 but, as discussed earlier in this report, the sampling methodology used was not sufficient. EPA has not collected new occurrence data on the contaminants since then. As a result, the older data may not provide an accurate basis for EPA to use when making regulatory determinations.

CONCLUSION

The overall goal of SDWA is to ensure that public drinking water is safe. The UCMR program is an important component of EPA's implementation of the act. As EPA's primary data source for assessing the nation's potential exposure to unregulated contaminants and the levels of that exposure, the UCMR plays a key role in EPA's regulatory development process. As a result, the accuracy and timeliness of the data it provides are important to support informed decision making. The technical changes EPA has made to UCMR3 in response to our 2011 recommendations are significant improvements over prior UCMR cycles. First, by more fully utilizing its statutory authority and monitoring for 30 contaminants under UCMR3, EPA has increased the productivity of the UCMR program. As a result, more data will be available to inform regulatory determinations in the future. Second, by increasing the sample size of the public water systems monitoring under the screening survey approach, EPA has also increased the statistical strength of its monitoring results. The larger sample size should allow EPA to make better estimates of exposure to drinking water contaminants. Third, by monitoring for contaminants at lower detection levels, or MRLs, the UCMR3 data are more likely to remain useful even in light of new information about the contaminants' health effects in the future.

However, some statutory provisions pose limitations for obtaining additional and more timely data. First, because SDWA caps the number of UCMR contaminants in each 5-year cycle at 30, the agency is not able to take advantage of economies of scale by collecting occurrence data on additional contaminants. As a result, EPA does not have the opportunity to collect data on additional unregulated contaminants at little additional effort and cost to the agency or the public water systems. The cost to detect additional contaminants with analytical methods already being used by the UCMR program can be nominal. Second, the 2-year window SDWA initially established from the time EPA publishes the UCMR list to when it publishes the regulatory determinations does not provide adequate time for the UCMR data to support regulatory determinations in the same cycle. As a practical matter, the time frames SDWA initially established are not adequate to provide enough time for UCMR's data collection period and EPA's subsequent use of the data. As a result, regulatory determinations on given contaminants are typically delayed by a full 5-year cycle, meaning that about 10 years pass from when EPA identifies them on the CCL as potential problems and decides whether they need to be regulated. Two contaminants—methyl bromide and 1,1-dichloroethane—will likely wait 22 years after their initial listing for regulatory determinations. Without congressional action to make the statutory time frames more realistic, the time lag and resultant delays are expected to continue in subsequent cycles.

The UCMR program faces other limitations that may cause EPA to base its regulatory determinations on inaccurate or outdated information. In particular, EPA's standard monitoring frequency may not be accurately estimating the occurrence of certain contaminants in the public's drinking water, particularly those that occur sporadically like viruses or seasonally like pesticides. By not varying the monitoring frequency for contaminants when appropriate, for example by increasing the frequency during the months when EPA expects viruses to be present, the UCMR data may not provide reliable estimates of contaminant occurrence. Similarly, by not remonitoring contaminants from prior UCMR cycles when appropriate, EPA may not have sufficient data to support its regulatory determinations. EPA continues to use data from past UCMR cycles to inform its regulatory determinations, but those data sets have potential limitations. For example, EPA previously determined that some of the data collected under UCMR1 were not sufficient to make national estimates of contaminant occurrence. Because EPA has not remonitored those contaminants, the agency still does not have sufficient UCMR data about them. Similarly, some of the UCMR1 data may be outdated due to changing industrial and agricultural chemical use and environmental conditions. As a result, the existing UCMR data for certain contaminants may no longer accurately reflect current drinking water conditions. We are encouraged by recent statements from EPA officials indicating that they will consider whether adequate occurrence data for a contaminant had already been collected under a prior UCMR when selecting contaminants in the future. Finally, the timing of the UCMR program significantly reduces the availability of the data to support regulatory determinations. Specifically, the time lines for the contaminant selection process and monitoring period do not get the UCMR data into the hands of EPA officials in time to support regulatory determinations in the same cycle. The UCMR data are EPA's preferred source of information on contaminant occurrence because the program has been specifically designed to yield nationally representative estimates of drinking water contamination. However, if the UCMR data are not available, EPA either must use data from another source, or delay making regulatory determinations on specific contaminants. For most of the UCMR3 contaminants, over 10 years are likely to pass between the time that EPA first listed contaminants on CCL3 in 2009 and the time that regulatory determinations are made for those contaminants, expected in 2020. The SDWA deadlines, EPA's lengthy contaminant selection process, and its 3-year monitoring period all contribute to the time lag. As stated earlier, to change the statutory deadlines and adjust the time frames requires congressional action, and shortening the selection process and monitoring period requires EPA action. UCMR program officials coordinate closely with the other two EPA offices that make regulatory determinations and develop the CCL and are sensitive to the impact that timeliness has on each of these processes. We recognize that EPA officials must consider a variety of data quality objectives, logistical needs, and the burden on public water systems in managing the UCMR program, regulatory determinations, and CCL. But, since EPA officials consider the UCMR data to be their primary source of occurrence data for the regulatory determinations, it is important that these data also be timely. However, without a coordinated effort to address the time lag, most contaminants on the CCL will continue to wait 10 years or longer for a regulatory determination.

Matters for Congressional Consideration

- To take advantage of opportunities to collect UCMR data on additional unregulated contaminants, Congress should consider amending SDWA to give EPA the flexibility to select more than 30 contaminants for monitoring under the UCMR program if high-priority contaminants, such as those on the CCL or contaminants of emerging concern, can be included at minimal cost, with minimal additional burden on public water systems, and while using analytical methods that EPA is already employing.
- To optimize the ability of the UCMR data to support regulatory determinations, Congress should consider adjusting the statutory time frames for the UCMR and regulatory determinations cycles so that EPA can use the UCMR data to support regulatory determinations in the same cycle.

Recommendations for Executive Action

We recommend that the Administrator of the Environmental Protection Agency direct the Assistant Administrator of the Office of Water to take the following three actions:

- To help ensure that the UCMR data accurately reflect contaminant occurrence, vary the monitoring frequency for public water systems when the standard UCMR monitoring frequency is not expected to accurately detect the presence of contaminants, such as sporadically occurring viruses or pesticides that fluctuate seasonally.
- To help address potential limitations with the UCMR data, when considering if adequate data had been collected under a prior UCMR during the contaminant selection process, take into account whether:
    - the sampling methodology EPA used in a prior UCMR for a particular contaminant provided sufficient occurrence data to make a regulatory determination,

    - the contaminant's health information is in flux or has been updated and now indicates that adverse health effects occur at a lower level than that at which UCMR data were originally collected, or
    - the data collected in an earlier cycle have become outdated and no longer present an accurate picture of contaminant occurrence, for example, because industrial or agricultural chemical use has changed.
- To optimize the ability of the UCMR program to support regulatory determinations, convene an internal working group of officials responsible for the UCMR, regulatory determinations, and CCL to fully examine opportunities to improve the timeliness of the UCMR program. Among other things, the working group should consider:
    - whether a shorter process for selecting contaminants would yield a high-quality list, or whether the existing process can be started sooner, and
    - whether monitoring for contaminants over a shorter period, instead of the current 3-year period, is feasible, given data quality, logistical needs, and the burden on public water systems.

AGENCY COMMENTS

We provided EPA with a draft of this report for its review and comment. In written comments, EPA generally agreed with our findings and conclusions and is committed to implementing our recommendations when it develops the next UCMR.

EPA also provided technical comments, which we incorporated as appropriate.

J. Alfredo Gómez
Director, Natural Resources and Environment

APPENDIX I: OBJECTIVES, SCOPE, AND METHODOLOGY

This report addressed the following objectives (1) evaluate the extent to which the Environmental Protection Agency (EPA) implemented the recommendations GAO made in 2011 to improve the Unregulated Contaminant Monitoring Rule (UCMR) program and any opportunities that may exist to strengthen it further; (2) identify the factors EPA considered when it selected the contaminants for monitoring under UCMR3 and the limitations, if any, that EPA faced in selecting the contaminants; and (3) examine the extent to which UCMR data support EPA's regulatory determinations.

To obtain information for each of our objectives on issues related to monitoring unregulated drinking water contaminants, we surveyed a broad range of experts in two rounds using the Delphi method. The Delphi method consists of surveying experts in two or more rounds to obtain group agreement or consensus on key topics. For this report, the first round of the survey was comprised of open-ended questions that asked experts to provide their views on the strengths and weaknesses of EPA's UCMR program, as well as suggestions for improving the program. The second round included primarily closed-ended questions developed from our content analysis of the first round survey responses. To identify experts for the first round of the survey, we reviewed public comments provided on EPA's proposed rules for its Contaminant Candidate Lists (CCL), UCMR, and preliminary regulatory determinations. We also reviewed public nominations of drinking water contaminants submitted to EPA and EPA documentation on two expert panels it convened to provide input and review of chemical and microbial contaminants for CCL3. We identified 14 experts from academia, an analytical laboratory, environmental advocacy organizations, professional associations, water utilities, and federal and state government agencies. These experts were selected based on their demonstrated familiarity with EPA's CCL, UCMR, and regulatory determination processes. The first round of the survey— which ran from December 12, 2012 to January 4, 2013—asked these experts to respond to 10 open-ended questions. These questions asked about various aspects of the UCMR program that helped or hindered EPA in (1) selecting priority contaminants, (2) monitoring for selected contaminants, and (3) using the data to support regulatory determinations. We also asked the experts whether programmatic or statutory changes were needed to address any challenges identified. We received responses from 14 of the 14 experts we surveyed in the first round of the survey.

We performed a content analysis of the responses from the first round of the survey to identify the most important issues raised by the experts. We used the results of the content analysis to develop questions for the second round of the survey. Because this was not a sample survey, it had no sampling errors. However, the practical difficulties of conducting any survey can introduce nonsampling errors, such as difficulties interpreting a particular question, which can introduce unwanted variability into the survey results. We took steps to minimize nonsampling errors by pretesting the questionnaire with five experts. This helped ensure that the questions were clear and unbiased, and that the questionnaire did not place an undue burden on respondents. An independent reviewer within GAO also reviewed a draft of the questionnaire prior to its administration. We made appropriate revisions to the content and format of the second round survey questionnaire based on the pretests and independent review.

To identify experts for the second round of the survey, we developed a preliminary list of experts, which included the 14 experts who completed the first round of the survey and others identified through interviews with stakeholders and our review of public comments provided on EPA's proposed rules. We contracted with the National Academy of Sciences to identify additional experts. We selected 48 experts for the second round of the survey based on their knowledge of one or more of the following topic areas: drinking water, contaminant monitoring, analytical methods, toxicology, risk assessment, and environmental regulation. Similar to the first round of the survey, the selected experts were from academia, analytical laboratories, environmental advocacy organizations, industry, water utilities, and federal and state government agencies. The second round of the survey was administered on the Internet from May 10, 2013 to June 7, 2013. Experts were sent an e-mail invitation to complete the survey on a GAO web server using a unique username and password. To increase the response rate, we followed up with e-mails and personal phone calls to the experts to encourage participation in our survey. The second round of the survey had an 83 percent response rate. We received responses from 40 of 48 experts, including 12 who had also completed round one of the survey. The information that we obtained from the expert survey is not generalizable to all experts that have an interest in or knowledge of drinking water contaminant monitoring issues. The full survey and responses are available in appendix IV. The survey participants are listed in appendix V.

To evaluate the extent to which EPA implemented GAO's May 2011 recommendations to improve the UCMR program, we reviewed EPA's proposed and final rules for the UCMR program and interviewed the relevant EPA officials who are responsible for designing and implementing the UCMR program to identify specific actions the agency had taken. To identify any opportunities that may exist to strengthen the UCMR program further, we interviewed stakeholders to obtain their perspectives on the UCMR program and related issues. In addition, we assessed EPA's sampling methodology against standard statistical criteria to determine whether the resulting sample sizes are likely to provide EPA with sufficient data for making national estimates of contaminant occurrence and exposure within an acceptable margin of error.

To identify the factors EPA considered when it selected the contaminants for monitoring under UCMR3, we reviewed EPA documentation describing the steps the agency took to select the contaminants, including UCMR3 working group meeting minutes and EPA's proposed and final rules for the UCMR program and the CCL, which serves as EPA's primary basis for selecting UCMR contaminants. We also interviewed EPA officials who are responsible for designing and implementing the UCMR program, including selecting the UCMR contaminants; as well as those responsible for developing the CCL and making regulatory determinations. To identify the limitations, if any, EPA faced in selecting the contaminants, we interviewed stakeholders to obtain their perspectives on the UCMR contaminant selection process and related issues. In addition, we assessed the UCMR contaminant selection process against statutory requirements.

To examine the extent to which UCMR data support EPA's regulatory determinations, we reviewed EPA and Federal Register documents about the use of the data for making the CCL2 and CCL3 regulatory determinations. In addition, we interviewed EPA officials who are responsible for designing and implementing the UCMR program and for making regulatory determinations. We also considered the time frames for implementing the UCMR program and using UCMR data for making regulatory determinations and assessed them against statutory requirements and GAO criteria for program timeliness. There is broad recognition throughout the federal government that, in addition to being effective in producing desired outcomes and other important dimensions of performance, agencies need to operate programs and provide information about them in a timely and efficient manner. For example, the GPRA Modernization Act calls on agencies to set goals and measure their progress toward meeting them. Timeliness is an important dimension of program performance. In addition, EPA initiated a new drinking water strategy in 2010 that recognizes the need to improve the timeliness of its efforts to address unregulated contaminants.

We conducted this performance audit from June 2012 to January 2014 in accordance with generally accepted government auditing standards. Those standards require that we plan and perform the audit to obtain sufficient, appropriate evidence to provide a reasonable basis for our findings and conclusions based on our audit objectives. We believe that the evidence obtained provides a reasonable basis for our findings and conclusions based on our audit objectives.

APPENDIX II: UCMR3 CONTAMINANTS

Assessment monitoring	
1,4-dioxane	vanadium
molybdenum	strontium
cobalt	chromium 6 (hexavalent chromium)
1,2,3-trichloropropane	total chromium[a]
1,3-butadiene	chlorate
chloromethane (methyl chloride)	perfluorooctanesulfonic acid
1,1-dichloroethane	perfluorooctanoic acid
bromochloromethane (Halon 1011)	perfluorononanoic acid
bromomethane (methyl bromide)	perfluorohexanesulfonic acid
chlorodifluoromethane (HCFC-22)	perfluoroheptanoic acid
	perfluorobutanesulfonic acid
Screening survey	
17-β-estradiol	estriol
17-α-ethynylestradiol (ethinyl estradiol)	equilin
estrone	testosterone
4-androstrene-3,17-dione	
Pre-screen testing	
enteroviruses	noroviruses

Source: UCMR3 final rule.

[a]Total chromium is a regulated contaminant. EPA used a different statutory authority, section 1445(a)(1)(A) of the Safe Water Drinking Act, to require monitoring for total chromium.

APPENDIX III: INFORMATION ON ANALYTICAL METHODS

Analytical methods are laboratory tests that measure the level of chemical and microbial contamination in a sample of drinking water. Analytical methods specify the volume of the sample to be tested, laboratory equipment and procedure used to test the sample, and quality control procedures to ensure the test is run correctly. Analytical methods vary based on the type of contaminant; different equipment and techniques are required for analyzing chemicals versus microbes. For example, laboratories use specialized scientific equipment such as gas chromatograph-mass spectrometers (GC-MS) to detect the presence of some chemicals in drinking water. Because the Unregulated Contaminant Monitoring Rule (UCMR) collects nationwide data, the Environmental Protection Agency's (EPA) goal is to use analytical methods that offer consistent application across the country. To get the data quality and accuracy that EPA needs, the agency uses analytical methods for UCMR that, among other requirements (1) have sufficient "holding times" for shipment of water samples from water utilities to laboratories; (2) can be used by a variety of laboratories with varying levels of expertise; and (3) can be used equally effectively under different drinking water conditions and types.[1] EPA often develops analytical methods for the UCMR program that meet these specific requirements. When EPA develops an analytical method, there are many key steps it follows. EPA begins with optimizing instrumentation and determining the best calibration, which involves setting up the equipment to measure a contaminant at the desired level. Next, EPA staff optimize the solid phase extraction—a process that involves separating compounds that are dissolved or suspended in water from other compounds in the water sample according to their physical and chemical properties in order to isolate the contaminant for analysis. Then, EPA staff determine which preservatives to use that will allow the chemical to be preserved in the water sample. They also determine any interferences– if there are other chemicals or substances in the water sample that may have similar physical and chemical properties and thus interfere with the chemical to be analyzed—and how to address such interferences, if any. Then, EPA staff determine the holding times for the contaminant and the water sample; as noted earlier, EPA must have sufficient "holding times" for shipment of water samples from water utilities to laboratories. These holding times must not have any negative impact in quality on the contaminant to be analyzed in the water sample. After following all these steps, if the analytical method meets EPA's data quality objectives, then EPA writes the method so that laboratories across the country can use and follow the EPA instructions to measure the level of contamination in drinking water. EPA then sends the written analytical method to several laboratories to ensure the method works as written; this process is known as a multilab verification process. If the method passes this verification process, then EPA publishes and uses the analytical method in a future UCMR. On the other hand, if the analytical method does not meet the data quality objectives that EPA has set, EPA revises the technical approach in order for the earlier steps in the analysis to lead to a method that meets the data quality objectives. See figure 6.

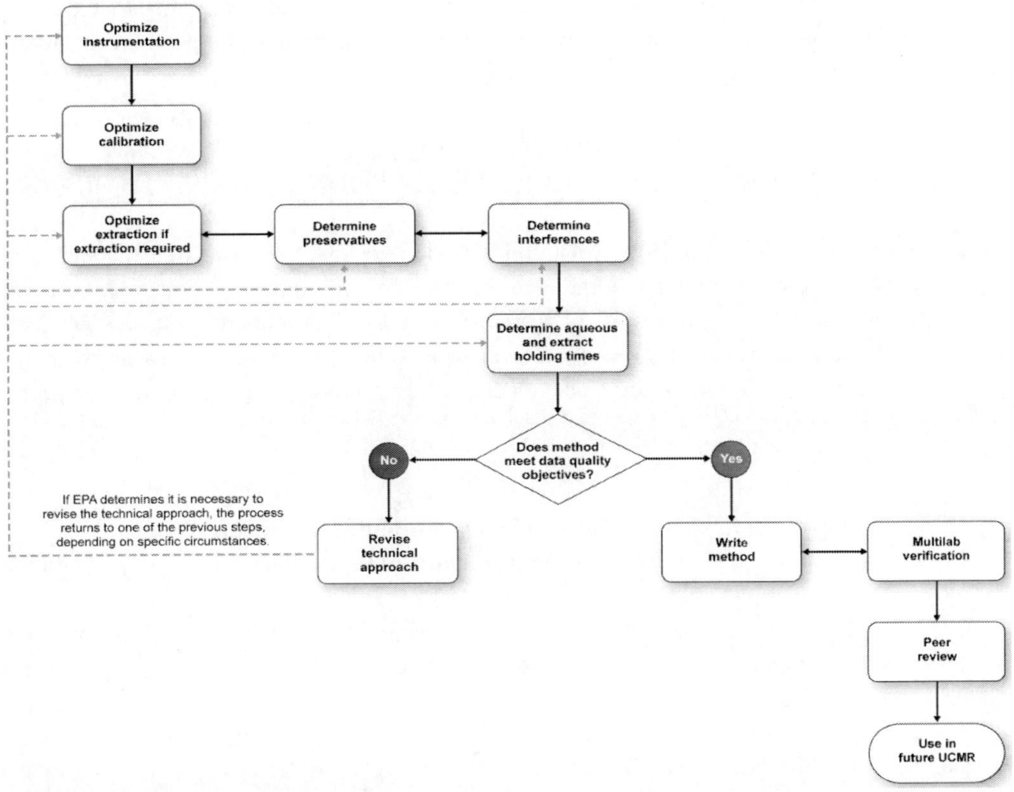

UCMR = Unregulated Contaminant Monitoring Rule.
Source: GAO analysis of EPA information.

Figure 6. EPA's Process for Analytical Method Development.

Developing an analytical method can be a complex, technically challenging, and time-consuming process. For example, EPA officials cited challenges for developing methods for some CCL3 contaminants such as the high water solubility of some contaminants, as well as some contaminants' instability in water. In such cases, EPA cannot keep these contaminants stable in the water samples for the minimum holding times that are required between the collection of samples by public water systems and sending the samples to the laboratories for analysis. For these contaminants, EPA is still working on techniques that will allow the agency to develop an analytical method that works effectively under all the conditions noted above that UCMR analytical methods must meet. EPA officials said it can take 1 to 2 years to develop a standard method that meets their data quality requirements and 2 to 3 years for more complex methods.

Most often, the UCMR program uses EPA analytical methods, but it also allowed some methods developed by others as alternatives for UCMR3. EPA is also using compliance methods in the UCMR3 program. Compliance methods are analytical methods that laboratories are required to use when monitoring for regulated drinking water contaminants. For example, the analytical method for volatile organic compounds that EPA is using for UCMR3 is a compliance method.[2] In addition, EPA involves stakeholders in its development of analytical methods. In May 2013, EPA held a public meeting and webinar to provide stakeholders with an update on analytical methods completed or under development for potential use in the UCMR program. Several external stakeholders participated in the meeting, both in person and via webinar. Some of them provided suggestions to address challenges EPA was facing with certain methods.

APPENDIX IV: ROUND 2 SURVEY QUESTIONNAIRE AND RESULTS

The questions we asked experts in Round 2 of our survey are shown below. Our survey was composed of closed- and open-ended questions. In this appendix, we include all the survey questions and aggregate results of responses to the closed-ended questions; we do not provide information on responses provided to the open-ended questions. For a more detailed discussion of our survey methodology, see appendix I. For a list of the experts who completed our survey, see appendix V.

Survey on EPA's Unregulated Contaminant Monitoring Program
U.S. Government Accountability Office

EPA uses the Unregulated Contaminant Monitoring Rule (UCMR) program to collect data for some contaminants that are suspected to be present in drinking water, but do not have health-based standards set under the Safe Drinking Water Act. Under the Safe Drinking Water Act, EPA is responsible for identifying and regulating drinking water contaminants that may pose a public health risk. Specifically, the act directs EPA to:

1) Publish a list every 5 years of contaminants that are a) not currently regulated b) known or anticipated to occur in public water systems and c) may require regulation. EPA has implemented this provision as the Contaminant Candidate List.
2) Establish a program to monitor unregulated contaminants and publish a list every 5 years of up to 30 contaminants to be monitored by public water systems. EPA has implemented this provision as the Unregulated Contaminant Monitoring Rule (UCMR) program.
3) Determine whether or not to regulate at least 5 contaminants included on the list of unregulated contaminants every 5 years. EPA has implemented this provision as the regulatory determination process. The act specifies that EPA shall regulate a drinking water contaminant under certain conditions. To see these conditions, click here. [In the online-version of the survey, this was a live link to the Safe Drinking Water Act requirements for making a regulatory determination.]

Some of the contaminants on the Contaminant Candidate List are monitored under the UCMR. EPA uses the monitoring data, along with information about the contaminants' health effects, to determine whether or not to regulate the contaminants. EPA is currently in its third round of the UCMR program (UCMR-3).

EPA is monitoring 30 contaminants under UCMR-3. Click here to review the list. [In the online-version of the survey this was a live link. See appendix II of this report for a list of the 30 contaminants.]

Section 1: UCMR Program Requirements and Considerations

The Safe Drinking Water Act requires that EPA publish a list of unregulated contaminants that pose the greatest public health concern, are known or anticipated to occur in public water systems, and may require regulation. This is known as the Contaminant Candidate List (CCL).

1. EPA uses the Unregulated Contaminant Monitoring Rule (UCMR) program to collect monitoring data for informing its regulatory determinations about some of the contaminants on the CCL. The Safe Drinking Water Act requires EPA to select contaminants that pose the greatest public health concern for its CCL and for regulatory determinations.

When selecting contaminants for monitoring under the UCMR, should EPA select contaminants that potentially pose the greatest public health concern?

Q1	Frequency	Cumulative Frequency
1-Definitely yes	24	24
2-Probably yes	13	37
3-Probably no	1	38
5-Not enough information to judge	2	40

2. EPA identified the contaminants for its third UCMR list (UCMR-3) largely from contaminants identified in its third Contaminant Candidate List (CCL-3). To create CCL-3, EPA identified about 7,500 contaminants and then analyzed the contaminants' potential to

occur in public water systems and to be a public health concern. Based on those criteria, EPA selected 116 contaminants for CCL-3 which it determined pose the greatest public health concern.

Was CCL-3 a reasonable source for EPA to use when it began identifying contaminants for the UCMR-3?

Q2	Frequency	Cumulative Frequency
1-Definitely yes	17	17
2-Probably yes	20	37
3-Probably no	2	39
5-Not enough information to judge	1	40

Please explain your answer to question 2: [Open ended]

3. The UCMR and CCL are created through separate processes. EPA finalized CCL-3 in 2009, and finalized UCMR-3 in 2012. UCMR-3 includes 22 contaminants from CCL-3 and 8 other contaminants.

If EPA combined the CCL and UCMR contaminant selection processes, would the resulting UCMR list likely be of similar or better quality when compared to the UCMR list created under the current separate processes?

Q3	Frequency	Cumulative Frequency
1-Definitely yes	5	5
2-Probably yes	15	20
3-Probably no	10	30
5-Not enough information to judge	10	40

Please explain your answer to question 3: [Open ended]

4. The Safe Drinking Water Act requires EPA to issue a new list of contaminants for monitoring under the UCMR every 5 years. Is 5 years enough time to select contaminants and implement a new round of the UCMR program?

Q4	Frequency	Cumulative Frequency
1-Definitely yes	10	10
2-Probably yes	18	28
3-Probably no	10	38
4-Definitely no	2	40

5. For UCMR-3, EPA selected 8 contaminants that were not on CCL-3. Some of these 8 contaminants became priorities between the publication of the CCL-3 and UCMR-3 lists.

Is it worthwhile for EPA to select contaminants for the UCMR when they are not on the CCL?

Q5	Frequency	Cumulative Frequency
1-Definitely yes	20	20
2-Probably yes	13	33
3-Probably no	7	40

6. The UCMR list of contaminants is based largely on the CCL, and UCMR supports regulatory determinations.

Do the 30 contaminants EPA selected warrant being included in UCMR-3? Click here to review the list. [In the online-version of the survey this was a live link. See appendix II of this report for a list of the 30 contaminants.]

Q6	Frequency	Cumulative Frequency
1-Definitely yes	11	11
2-Probably yes	19	30
3-Probably no	3	33
4-Definitely no	3	36
5-Not enough information to judge	1	37
6-No answer	3	40

7. EPA added chromium-6 and total chromium to the UCMR-3 list based largely on the strength of public comments supporting these additions.

7A. Did chromium-6 warrant being included on the UCMR-3 list?

Q7A	Frequency	Cumulative Frequency
1-Definitely yes	19	19
2-Probably yes	16	35
3-Probably no	3	38
5-Not enough information to judge	2	40

7B. Did total chromium warrant being included on the UCMR-3 list?

Q7B	Frequency	Cumulative Frequency
1-Definitely yes	11	11
2-Probably yes	17	28
3-Probably no	8	36
5-Not enough information to judge	4	40

Section 2: Factors EPA Considers when Selecting Contaminants for UCMR

8. One factor EPA considers when selecting contaminants for UCMR is what health effects information is available for the universe of unregulated contaminants.

In general, was enough health effects information available when EPA began selecting contaminants in 2009 to identify contaminants that potentially pose the greatest public health concern?

Q8	Frequency	Cumulative Frequency
1-Definitely yes	2	2
2-Probably yes	14	16
3-Probably no	15	31
4-Definitely no	2	33
5-Not enough information to judge	4	37
6-No answer	3	40

9. EPA also needs health effects information when determining whether or not to regulate a contaminant monitored under UCMR. However, health effects information is often being developed for contaminants at the same time that the UCMR program is collecting data about contaminant occurrence.

Given this situation, is it worthwhile for EPA to select contaminants for UCMR when the health effects information needed to make regulatory determinations is not yet available?

Q9	Frequency	Cumulative Frequency
1-Definitely yes	8	8
2-Probably yes	21	29
3-Probably no	8	37
4-Definitely no	2	39
6-No answer	3	40

10. Another factor EPA considers when selecting contaminants for UCMR is whether or not the contaminant was included in a prior round of UCMR. EPA typically does not select contaminants that were previously included. While this policy reduces the burden on public water systems, there have been situations in which monitoring again could enable EPA to collect better data. The following questions ask about such situations.

10A. In the first round of the UCMR program, the sampling methodology that EPA used to select water systems had limitations, and therefore the program did not provide sufficient data for making decisions about whether or not to regulate some of the contaminants.

Given the tradeoff between burden on water systems and data quality, would it be worthwhile for EPA to consider selecting some of those contaminants again for monitoring under UCMR?

Q10A	Frequency	Cumulative Frequency
1-Definitely yes	15	15
2-Probably yes	16	31
3-Probably no	7	38
4-Definitely no	1	39
5-Not enough information to judge	1	40

10B. In prior rounds of the UCMR program, some of the analytical methods used were not sensitive enough to detect contaminants at EPA's health effects benchmark. More sensitive analytical methods have since become available for some of these contaminants.

Given the tradeoff between burden on water systems and data quality, would it be worthwhile for EPA to consider selecting some of those contaminants again for monitoring under UCMR?

Q10B	Freque	Cumulative Frequency
1-Definitely yes	18	18
2-Probably yes	16	34
3-Probably no	3	37
5-Not enough information to	3	40

10C. New or updated health effects information is now available for some contaminants monitored in prior rounds of the UCMR. The new information indicates that health effects occur at lower concentrations than those previously monitored for.

Given the tradeoff between burden on water systems and data quality, would it be worthwhile for EPA to consider selecting some of those contaminants again for monitoring under UCMR?

Q10C	Frequency	Cumulative Frequency
1-Definitely yes	19	19
2-Probably yes	13	32
3-Probably no	4	36
5-Not enough information to judge	4	40

11. Another factor EPA considers when selecting contaminants for UCMR is whether analytical methods will be available to detect these contaminants in drinking water. EPA typically develops analytical methods for the contaminants it plans to monitor. For UCMR-3, EPA also considered analytical methods that had been developed by Standard Methods and the American Society for Testing and Materials.

For future rounds of UCMR, will it be worthwhile for EPA to reach out to organizations that develop such analytical methods before it selects contaminants?

Q11	Frequency	Cumulative Frequency
1-Definitely yes	32	32
2-Probably yes	8	40

12. Another factor EPA considers when selecting contaminants for UCMR is whether the contaminants are similar to others it will be monitoring. EPA selected 6 contaminants for UCMR-3 primarily because they could be detected by an analytical method that will be used for another UCMR-3 contaminant. These 6 were not on the CCL-3, EPA's primary source for selecting UCMR-3 contaminants.

Is it worthwhile for EPA to select contaminants for the UCMR primarily because they can be detected by an analytical method that will be used for another UCMR contaminant?

Q12	Frequency	Cumulative Frequency
1-Definitely yes	10	10
2-Probably yes	17	27
3-Probably no	12	39
4-Definitely no	1	40

13. Another factor EPA considers when selecting contaminants is stakeholder input. For UCMR-3, EPA convened a State working group, held a public stakeholder meeting, and collected public comments on the proposed rule published in the Federal Register.

Did these means of communication give stakeholders enough opportunity to provide input on the contaminants that EPA selected for UCMR-3?

Q13	Frequency	Cumulative Frequency
1-Definitely yes	10	10
2-Probably yes	23	33
3-Probably no	5	38
4-Definitely no	1	39
5-Not enough information to judge	1	40

14. When EPA develops the CCL it is reviewed by EPA's Science Advisory Board. In addition, EPA solicited additional expert input from the National Drinking Water Advisory Board and from other outside experts when it selected contaminants for the CCL-3.

Should EPA use similar means to obtain additional expert input when it selects contaminants for the UCMR?

Q14	Frequency	Cumulative Frequency
1-Definitely yes	24	24
2-Probably yes	15	39
3-Probably no	1	40

15. In addition to EPA, the US Geological Survey, some states, and other groups generate contaminant occurrence data. EPA reviews some of this other data when selecting the UCMR contaminants.

Is it worthwhile for EPA to consider this other data on contaminant occurrence when it selects contaminants for UCMR?

Q15	Frequency	Cumulative Frequency
1-Definitely yes	32	32
2-Probably yes	7	39
3-Probably no	1	40

Section 3: Collection of Contaminant Occurrence Data through the UCMR Program

16. EPA varies the frequency for monitoring based primarily on the source water type. Under UCMR-3, **surface water systems** must monitor for **4** consecutive quarters with sampling events 3 months apart.

Is the requirement that **surface water systems** monitor for **4** consecutive quarters likely to provide EPA sufficient data to determine whether contaminants occur with a frequency and at levels of public health concern?

Q16	Frequency	Cumulative Frequency
1-Definitely yes	1	1
2-Probably yes	26	27
3-Probably no	7	34
4-Definitely no	5	39
5-Not enough information to judge	1	40

Please explain your answer to question 16: [Open ended]

17. Under UCMR-3, **ground water systems** must monitor **twice** in a 12-month period with sampling events 5- 7 months apart.

Is the requirement that **ground water systems** monitor **twice** in a 12-month period likely to provide EPA sufficient data to determine whether contaminants occur with a frequency and at levels of public health concern?

Q17	Frequency	Cumulative Frequency
1-Definitely yes	3	3
2-Probably yes	25	28
3-Probably no	10	38
4-Definitely no	2	40

Please explain your answer to question 17: [Open ended]

18. Under UCMR-3, **non-disinfecting ground water systems** located in sensitive hydrogeological areas (i.e., karst or fractured bedrock) must monitor **twice** in a 12-month period with sampling events 5-7 months apart for 2 viruses.

Is the requirement that these **non-disinfecting ground water systems** monitor **twice** in a 12-month period likely to provide EPA sufficient data to determine whether the contaminants occur with a frequency and at levels of public health concern?

Q18	Frequency	Cumulative Frequency
2-Probably yes	9	9
3-Probably no	16	25
4-Definitely no	7	32
5-Not enough information to judge	4	36
6-No answer	4	40

Please explain your answer to question 18: [Open ended]

19. UCMR-3 is a national study with samples taken over 3 years (January 2013 through December 2015) and across all months and seasons to capture seasonal fluctuations in contaminant occurrence.

19A. For the 30 contaminants being monitored under UCMR-3, is this **3-year** monitoring timeframe likely to provide EPA sufficient information on seasonal fluctuations in contaminant occurrence? Click here to review the list of 30 contaminants. [In the online-version of the survey this was a live link. See appendix II of this report for a list of the 30 contaminants.]

Q19A	Frequency	Cumulative Frequency
1-Definitely yes	5	5
2-Probably yes	27	32
3-Probably no	4	36
4-Definitely no	1	37
5-Not enough information to judge	3	40

19B. For the 30 contaminants being monitored under UCMR-3, would a **2-year** monitoring timeframe instead of the **3-year** timeframe be likely to provide EPA sufficient information on seasonal fluctuations in contaminant occurrence?

Q19B	Frequency	Cumulative Frequency
1-Definitely yes	3	3
2-Probably yes	16	19
3-Probably no	9	28
4-Definitely no	2	30
5-Not enough information to judge	9	39
6 – No answer	10	40

20. EPA established a three-tiered strategy for monitoring UCMR contaminants based primarily on the availability of analytical methods. UCMR's Tier I monitoring requires all water systems serving more than 10,000 people and a random sample of 800 water systems serving 10,000 or fewer people to monitor those contaminants that can be detected by analytical methods that are widely used in drinking water laboratories.

Is Tier I monitoring likely to provide EPA sufficient data to determine whether contaminants occur with a frequency and at levels of public health concern?

Q20	Frequency	Cumulative Frequency
1-Definitely yes	4	4
2-Probably yes	27	31
3-Probably no	3	34
5-Not enough information to judge	4	38
6-No answer	2	40

21. UCMR's Tier II monitoring requires all water systems serving more than 100,000 people, 320 randomly selected water systems serving 10,001 to 100,000 people, and 480

randomly selected water systems serving 10,000 or fewer people to monitor those contaminants that can be detected by newly developed analytical methods that are used in fewer drinking water laboratories.

Is Tier II monitoring likely to provide EPA sufficient data to determine whether contaminants occur with a frequency and at levels of public health concern?

Q21	Frequency	Cumulative Frequency
1-Definitely yes	1	1
2-Probably yes	25	26
3-Probably no	8	34
5-Not enough information to judge	4	38
6-No answer	2	40

22. UCMR's Tier III monitoring requires a random sample of 800 non-disinfecting ground water systems located in sensitive hydrogeological areas (i.e., karst or fractured bedrock) to monitor those contaminants that can be detected by very new or specialized analytical methods that are used in a small set of drinking water laboratories.

Is Tier III monitoring likely to provide EPA sufficient data to determine whether contaminants occur with a frequency and at levels of public health concern?

Q22	Frequency	Cumulative Frequency
1-Definitely yes	1	1
2-Probably yes	17	18
3-Probably no	11	29
4-Definitely no	2	31
5-Not enough information to judge	8	39
6-No answer	1	40

23. Overall, is the three-tiered monitoring strategy described in questions 20-22, which varies the number and size of water systems based on the availability of analytical methods, likely to provide EPA sufficient data to determine whether contaminants occur with a frequency and at levels of public health concern?

Q23	Frequency	Cumulative Frequency
2-Probably yes	26	26
3-Probably no	5	31
4-Definitely no	1	32
5-Not enough information to judge	6	38
6-No answer	2	40

Please explain your answer to question 23: [Open ended]

24. Under UCMR-3, systems must monitor for 5 pathogen indicators. This monitoring will be conducted in conjunction with virus monitoring of enterovirus and norovirus.

The pathogen indicators are: Total coliforms, *E. coli*, *Enterococci,* Bacteriophage, and Aerobic spores.

Is this type of monitoring likely to help EPA better understand the co-occurrence of pathogen indicators and viruses?

Q24	Frequency	Cumulative Frequency
1-Definitely yes	9	9
2-Probably yes	21	30
3-Probably no	3	33
4-Definitely no	1	34
5-Not enough information to judge	3	37
6-No answer	3	40

25. Under UCMR-3, systems must monitor for total chromium in conjunction with chromium-6.

25A. Is this type of monitoring likely to provide EPA with valuable information on the co-occurrence of total chromium and chromium-6?

Q25A	Frequency	Cumulative Frequency
1-Definitely yes	14	14
2-Probably yes	20	34
3-Probably no	2	36
5-Not enough information to judge	3	39
6-No answer	1	40

25B. Is this type of monitoring likely to provide EPA with valuable information on the usefulness of monitoring for total chromium as an indicator for chromium-6?

Q25B	Frequency	Cumulative Frequency
1-Definitely yes	7	7
2-Probably yes	19	26
3-Probably no	8	34
4-Definitely no	2	36
5-Not enough information to judge	3	39
6-No answer	1	40

Section 4: Use of the UCMR Contaminant Occurrence Data

26. EPA uses UCMR data to make regulatory determinations for some of the contaminants listed on the CCL. Due to a time lag, the UCMR-3 occurrence data will not be available to inform the regulatory determinations about the contaminants on CCL-3. As a result, EPA is using occurrence data from prior UCMRs to inform these regulatory determinations. CCL-3 includes 28 contaminants monitored under prior UCMRs.

Is this likely to be an effective strategy for making regulatory determinations, given the time lag?

Q26	Frequency	Cumulative Frequency
1-Definitely yes	1	1
2-Probably yes	9	10
3-Probably no	18	28
4-Definitely no	7	35
5-Not enough information to judge	4	39
6-No answer	1	40

Please explain your answer to question 26: [Open ended]

27. Sometimes EPA waits several years before it uses the UCMR occurrence data to inform its regulatory determinations.

After the UCMR occurrence data has been collected, about how long is it likely to be a reliable indicator of the frequency and level of drinking water contamination?

Q27	Frequency	Cumulative Frequency
b.-Less than 5 years	11	11
c.-5 to less than 11 years	8	19
d.-11 to less than 16 years	2	21
e.-16 to less than 20 years	1	22
g.-Not enough information to judge	18	40

28. The primary purpose of the UCMR program is to collect contaminant occurrence data for informing regulatory determinations which are based, in part, on the frequency and level of drinking water contamination.

Overall, how effective has the UCMR program been at collecting sufficient data to determine whether contaminants occur with a frequency and at levels of public health concern?

Q28	Frequency	Cumulative Frequency
1-Very effective	9	9
2-Somewhat effective	26	35
3-Not effective	1	36
4-Not enough information to judge	3	39
5-No answer	1	40

29. Another purpose of the UCMR program is to inform EPA's development of future CCLs. EPA expects to complete its data collection and analysis for UCMR-3 in 2016; it expects to finalize its next CCL in 2014.

Given the tradeoff between data completeness and timeliness, would it be worthwhile for EPA to use partial UCMR data to inform its next CCL?

Q29	Frequency	Cumulative Frequency
1-Definitely yes	11	11
2-Probably yes	24	35
3-Probably no	5	40

30. An additional purpose of the UCMR program is to inform EPA's health research planning. For example, EPA might prioritize the completion of health effects information for a specific contaminant over others if it believes the contaminant is likely to be found in drinking water. EPA expects to complete its data collection and analysis for UCMR-3 in 2016; its health research planning is on-going.

Given the tradeoff between data completeness and timeliness, would it be worthwhile for EPA to use partial UCMR data to inform its health research planning?

Q30	Frequency	Cumulative Frequency
1-Definitely yes	16	16
2-Probably yes	20	36
3-Probably no	2	38
5-Not enough information to judge	2	40

IMPORTANT: Please continue to the next screen to submit your final responses to GAO.
Submit Your Final Responses to GAO

31. Are you ready to submit your final completed survey to GAO?

(This is equivalent to mailing a completed paper survey to us. It tells us that your answers are official and final.)

- Yes, my survey is complete - To submit your final responses, please click on "Exit" below
- No, my survey is not yet complete - To save your responses for later, please click on "Exit" below

APPENDIX V: LIST OF EXPERTS WHO COMPLETED THE SURVEYS

Scott Biernat, Association of Metropolitan Water Agencies
Melissa Billman, Fairfax Water Utility
Danielle Blacet, Association of California Water Agencies
Mark Borchardt, Ph.D., U.S. Department of Agriculture
Bryan Brooks, Ph.D., Baylor University
Gary Burlingame, Philadelphia Water Department
Peter L. deFur, Ph.D., Environmental Stewardship Concepts
Gregory Delzer, U.S. Geological Survey
Andy Eaton, Ph.D., Eurofins Eaton Analytical, Inc.
Stephen Estes-Smargiassi, Massachusetts Water Resources Authority
Mike Focazio, Ph.D., U.S. Geological Survey

David Fischer, American Chemistry Council
Ann Marie Gebhart, Ph.D., ToxServices LLC
Will Gonzalez, Coachella Valley Water District
Annette Guiseppi-Elie, Ph.D., DuPont
George Hallberg, Ph.D., The Cadmus Group, Inc.
Charles Hertz, Ph.D.,Aqua America, Inc.
Michael Hutcheson, Ph.D., Massachusetts Department of Environmental Protection
Elston Johnson, Texas Commission on Environmental Quality
Todd Johnson, Minnesota Department of Health
Nancy Kim, Ph.D., State University of New York at Albany
Rebecca D. Klaper, Ph.D., University of Wisconsin-Milwaukee
Dana Kolpin, U.S. Geological Survey
John H. Koon, Ph.D., Georgia Technical Institute
Mark LeChevallier, Ph.D., American Water
Gary R. Lynch, Park Water Company
Charles Maddox, Austin Water Utility
David Mazzera, Ph.D., California Department of Public Health
Erik Olson, Pew Charitable Trusts
Darrell Osterhoudt, Association of State Drinking Water Administrators
Rebecca Parkin, Ph.D., George Washington University
Gloria Post, Ph.D., New Jersey Department of Environmental Protection
Alan Roberson, American Water Works Association
Renee Sharp, Environmental Working Group
Mary Skopec, Ph.D., Iowa Department of Natural Resources
Shane Snyder, Ph.D., University of Arizona
Mark D. Sobsey, Ph.D., University of North Carolina at Chapel Hill
Lynn Thorp, Clean Water Action
Patricia Toccalino, Ph.D., U.S. Geological Survey
Lloyd Wilson, Ph.D., New York State Department of Health
Mae Wu, Natural Resources Defense Council
Nira Yamachika, Orange County Water District

End Notes

[1] Pub. L. No. 93-523 (1974), codified as amended at 42 U.S.C. §§300f-300j-26 (2013).

[2] Under SDWA, a drinking water contaminant is defined as any physical, chemical, biological, or radiological substance or matter in water.

[3] In the 1996 amendments, Congress removed a previous requirement that EPA regulate 25 new contaminants every 3 years.

[4] EPA decided to regulate perchlorate in 2011 and is currently developing a regulatory proposal for it. Perchlorate is both a naturally occurring and man-made chemical that is used to produce rocket fuel, fireworks, flares and explosives. It can also be present in bleach and in some fertilizers. According to EPA, perchlorate may have adverse health effects because scientific research indicates that this contaminant can disrupt the thyroid's ability to produce hormones needed for normal growth and development.

[5] The current Contaminant Candidate List is referred to as CCL3.

[6] For purposes of the UCMR, EPA defines a small water system as one serving 10,000 or fewer people. EPA defines a large water system as one that serves more than 10,000 people.

[7] GAO, *Safe Drinking Water Act: EPA Should Improve Implementation of Requirements on Whether to Regulate Additional Contaminants,* GAO-11-254 (Washington, D.C.: May 27, 2011).

[8] The Delphi method consists of surveying experts in two or more rounds to obtain group agreement or consensus on key topics.

[9] To protect drinking water from disease-causing organisms, or pathogens, water suppliers may add a disinfectant, such as chlorine, to drinking water. However, disinfectants can react with naturally occurring materials in the water to form by-products, some of which may pose health risks.

[10] States may regulate some of these contaminants. Under state laws, some state environmental agencies have the authority to regulate additional contaminants or establish more stringent standards than federal regulations, while others do not have these authorities.

[11] GAO-11-254.

[12] EPA defines a community water system as a public water system that has at least 15 service connections used by year-round residents or regularly serves an average of at least 25 year-round residents. EPA defines a nontransient, noncommunity water system as a public water system that is not a community water system and regularly serves at least 25 of the same people over 6 months per year. There are more than 52,000 community water systems and more than 19,000 nontransient, noncommunity water systems.

[13] EPA defines a transient noncommunity water system as a public water system that does not regularly serve at least 25 of the same people over 6 months per year. There are more than 87,000 transient, noncommunity water systems.

[14] 42 U.S.C. § 300g-1(b)(1)(C) (2013).

[15] Under the time frames set out in the statute, EPA would have issued CCL3 in 2008, the list of UCMR3 contaminants in 2009, and the regulatory determinations for CCL3 in 2011.

[16] EPA has issued or updated health advisories for the following 9 contaminants: (1) 1,1,2,2-tetrachloroethane; (2) 2,4-dinitrotoluene; (3) 2,6-dinitrotoluene; (4) boron; (5) tetrachloroterephthalic acid; (6) monomethyl tetrachloroterephthalic acid; (7) manganese; (8) sodium; and (9) sulfate. EPA also developed consumer information for *Acanthamoeba*.

[17] Laboratories conducting the pre-screen testing analyses for viruses and pathogen indicators are under direct contracts with EPA and are subject to an analogous laboratory approval process.

[18] These data are included in EPA's National Contaminant Occurrence Database. EPA developed the database to satisfy the statutory requirements set by Congress in the 1996 amendments to SDWA to maintain a national drinking water contaminant occurrence database using sampling data for both regulated and unregulated contaminants in public water systems.

[19] Holding time is the time between the collection of a water sample by a public water system and the laboratory analysis of that water sample.

[20] Three of the 17 recommendations we made in GAO-11-254 applied to the UCMR program. Other recommendations concern regulatory determinations and could affect the UCMR program due to the relationship between the programs.

[21] MRLs represent an estimate of the lowest concentration of a compound that can be quantitatively measured by members of a group of experienced drinking water laboratories. Under the UCMR, laboratories are to report all analyses of listed contaminants. Laboratories report values when they are equal to or greater than the established MRL, or report that the analysis was below the MRL.

[22] GAO-11-254.

[23]o ne of the 30 contaminants to be monitored under UCMR3 is total chromium, a regulated contaminant. EPA included total chromium in UCMR3 to collect data on it concurrent with data for chromium-6, an associated unregulated contaminant. EPA used a different statutory authority, section 1445(a)(1)(A) of SDWA, to require monitoring for total chromium.

[24] The 9 contaminants include 4 monitored under, or in association with UCMR1, and 5 monitored under the program that preceded the UCMR. Under the earlier program, public water systems were required to monitor for up to 48 unregulated contaminants and to report monitoring results to the states or to EPA.

[25] GAO-11-254.

[26] EPA set most of the MRLs for the UCMR3 contaminants at the level of micrograms per liter (μg/L), generally ranging from 0.0001 μg/L to 1 μg/L. Some contaminants are able to be detected at much lower concentrations than others, due to their chemical properties, among other factors. The levels set for the UCMR3 contaminants

are generally more sensitive than the levels set for UCMR1 and UCMR2 contaminants, which generally ranged from 0.5 μg/L to 10 μg/L and from 0.002 μg/L to 2 μg/L, respectively.

27 Groundwater under the direct influence of surface water means any water beneath the surface of the ground with significant occurrence of insects or other macroorganisms, algae, or large-diameter pathogens such as *Giardia lamblia* or *Cryptosporidium,* or significant and relatively rapid shifts in water characteristics such as turbidity, temperature, conductivity, or pH that closely correlate to climatological or surface water conditions. For example, a groundwater source located close enough to nearby surface water, such as a river or lake, to receive direct surface water recharge is considered at risk of contamination from pathogens that are not normally found in true groundwaters.

28 Environmental Protection Agency, "Revisions to the Unregulated Contaminant Monitoring Regulation for Public Water Systems," 64 Fed. Reg. 50,556, 50,570, 50,588 (Sept. 17, 1999).

29 Id. at 50,571 ("For much of the United States, east of the Rocky Mountains, many studies have shown that the season of greatest vulnerability for contaminant occurrence is the late-spring, early-summer runoff-recharge period, particularly for contaminants such as pesticides and nitrate (e.g., Larson et al., 1997; Barbash and Resek, 1996; Hallberg, 1989a,b). For deeper, more confined ground water systems, defining vulnerable periods is much more difficult.").

30 Systems that are unable to monitor within the assigned time frame must contact EPA to reschedule their monitoring.

31 Environmental Protection Agency, *UCMR2 Categorized Public Comments* (Washington, D.C.: June 2006); and Environmental Protection Agency, *Response to Comments Document for the Unregulated Contaminant Monitoring Regulations (UCMR3)* (Washington, D.C.: January 2012).

32 The 800 systems use groundwater wells that do not treat the water and are located in areas of vulnerable geology, such as karst aquifers.

33 According to an EPA fact sheet, the primary means of virus-related disease transmission is exposure to fecally contaminated water or consumption of food exposed to the contaminated water. Exposure to enterovirus-contaminated drinking water can cause fever, respiratory illness, rash, conjunctivitis, hand-foot-mouth disease, and meningitis, among other things. Exposure to norovirus-contaminated drinking water can cause mild fever, vomiting, diarrhea, and cramps.

34 *Wisconsin Water and Health Trial for Enteric Risks (WAHTER Study)—Part 1: Risk of Illness from Municipal Groundwater Consumption.* EPA Office of Water Symposium on Groundwater-Borne Infectious Disease, Etiologic Agents and Indicators (Washington, D.C.: December 2–4, 2008).

35 Environmental Protection Agency, *Public Comment and Response Summary for the Unregulated Contaminant Monitoring Regulation* (Washington, D.C.: August 1999).

36 According to EPA's website, *Aeromonas* is a genus of bacteria with the potential to contaminate drinking water by evading treatment (e.g., chlorine disinfection) and to cause human disease such as gastroenteritis and bacterial septicemia, which are typically acquired through ingestion or though exposure of open wounds.

37 Specifically, the three herbicides included in UCMR2 are acetochlor, alachlor, and metolachlor. Such pesticides have the potential to contaminate drinking water by making their way into surface water or groundwater systems that feed drinking water supplies, for example, through runoff or drainage resulting from rainfall or irrigation. EPA classified all three contaminants as "possible" or "probable" carcinogens. The six degradates included in UCMR2 are acetochlor ethane sulfonic acid; acetochlor oxanilic acid; alachlor ethane sulfonic acid; alachlor oxanilic acid; metolachlor ethane sulfonic acid; and metolachlor oxanilic acid. Degradates are the result of pesticides breaking down in the environment as they interact with water, soil, air, sunlight, or microorganisms.

38 The USGS National Water-Quality Assessment Program (NAWQA) conducted a national-scale assessment of pesticide occurrence and concentrations in streams and groundwater during 1992–2001. The assessment's findings were published in a March 2006 report entitled, *The Quality of Our Nation's Waters—Pesticides in the Nation's Streams and Ground Water, 1992–2001.* The report notes that "although NAWQA did not measure pesticide concentrations at drinking-water intakes, NAWQA results for the wide range of streams sampled indicate that seasonal pulses of pesticide concentrations probably occur in some streams that are used as sources of drinking water."

39 A contaminant of emerging concern is a contaminant characterized by a perceived, potential or real threat to human health or the environment or by lack of published health standards.

40 Environmental Protection Agency, "Revisions to the Unregulated Contaminant Monitoring Regulation (UCMR 3) for Public Water Systems," Proposed Rule, 76 Fed. Reg. 11,713 (Mar. 3, 2011).

[41] EPA's Office of Science and Technology, part of the Office of Water, provides scientific support for drinking water regulatory determinations, information for drinking water health advisories, and other human health risk assessments, among other duties.

[42] Total chromium, while already regulated in drinking water, was included in conjunction with UCMR3 monitoring to determine its relationship with chromium-6, a contaminant of emerging concern.

[43] EPA classified the UCMR program as a Tier 3 action under its Action Development Process. Tier 3 actions allow the lead office to design their own review processes and do not require formal EPA management reviews or OMB reviews.

[44] Given continuing budget pressures combined with the focus on performance envisioned in the GPRA Modernization Act of 2010, we have reported that federal agencies need to identify ways to operate more efficiently. We have defined efficiency as maintaining federal government services or outcomes using fewer resources (such as time and money) or improving or increasing the quality or quantity of services or outcomes while maintaining (or reducing) resources. See GAO, *Streamlining Government: Key Practices from Select Efficiency Initiatives Should be Shared Governmentwide*, GAO-11-908 (Washington, D.C.: Sept. 30, 2011).

[45] PFCs are man-made chemicals that are used in a wide variety of industrial and commercial products. EPA has reported that PFCs have significant adverse effects on laboratory animals and wildlife and anticipates that continued exposure to PFCs could result in adverse effects on humans.

[46] Environmental Protection Agency, *Long-Chain Perfluorinated Chemical's (PFCs) Action Plan* (Washington, D.C.: Dec. 30, 2009).

[47] EPA has funded the small public water system's UCMR monitoring from EPA's State Revolving Fund appropriation.

[48] EPA has only once remonitored for a contaminant: namely, the pesticide acetochlor, which was monitored under UCMR1 and UCMR2. MTBE was monitored under UCMR1 and was considered for, but not included, in UCMR3. EPA's practice when selecting contaminants for UCMR3 was to exclude those monitored under a prior UCMR, as stated in the *Federal Register*.

[49] Environmental Protection Agency, "Drinking Water: Regulatory Determinations Regarding Contaminants on the Second Drinking Water Contaminant Candidate List," Final Rule, 73 Fed. Reg. 44,251 (July 30, 2008).

[50] Environmental Protection Agency, "Revisions to the Unregulated Contaminant Monitoring Regulation for Public Water Systems," Final Rule, 64 Fed. Reg. 50,556, 50,568 (Sept. 17, 1999).

[51] The 300 public water systems included 120 systems serving more than 10,000 people and 180 systems serving 10,000 or fewer people.

[52] The 800 public water systems included 320 systems serving 10,001 to 100,000 people; and a representative sample of 480 systems serving 10,000 or fewer people. Approximately 400 systems serve more than 100,000 people.

[53] In 2009, EPA cancelled the registration for disulfoton. In 2004, EPA cancelled certain uses of molinate. Molinate is still registered but, as of 2012, there were no active uses, according to EPA documents.

[54] EPA's data do not indicate whether the decrease in nitrobenzene use occurred before, during, or after the UCMR1 data were collected.

[55] In May 2011, we reported that EPA had not defined what data are adequate to support a regulatory determination. In that report, we also recommended that EPA develop guidance for making regulatory determinations that included factors for determining when data provide adequate support. In its response to the recommendation, EPA stated that it did not believe establishing such guidance was practicable, given the many combinations of health effects factors and potential ranges of frequencies and levels of contaminants measured in drinking water.

[56] SDWA requires EPA, in carrying out the provisions concerning listing, selecting, and regulating contaminants, to the degree that an action is based on science, to use the best available peer-reviewed science and data collected by accepted or best available methods.

[57] SDWA requires EPA to make at least 5 regulatory determinations every 5 years, and hence we refer to the 5-year mark as a "deadline." See 42 U.S.C. § 300g-1(b)(1)(B)(ii)(I) (2013).

[58] EPA has made 20 negative regulatory determinations and 1 positive regulatory determination (perchlorate) since the SDWA Amendments of 1996 established the UCMR program.

[59] EPA also needs information on a contaminant's health effects to make a regulatory determination. In 2011, GAO reported that EPA did not have health effects information for many of the contaminants on the CCL, which limited EPA's ability to make regulatory determinations. See GAO-11-254.

[60] In its comments on a draft of this report, EPA noted that the agency makes judgments on a case by case basis to determine whether data are too old.

[61] SDWA limits EPA to listing no more than 30 contaminants every 5 years and, as a result, EPA cannot accelerate the UCMR schedule by issuing the next list earlier–that is, before 5 years from the last one–under current law. Each time EPA publishes a UCMR list or set of regulatory determinations, the agency adds 5 years to determine its next deadline.

[62] EPA had considered shortening the time to report, review, and approve the data by about half (90 days total) but decided not to, based on public comments on the proposed UCMR3 rule and OMB's review of the final rule.

[63] Final regulatory determinations for the CCL3 contaminants were originally expected to be published in 2013, according to EPA documents. However, EPA officials told us they have been delayed until 2015 for reasons unrelated to the UCMR program.

[64] Survey on EPA's Unregulated Contaminant Monitoring Program, Question 19B: For the 30 contaminants being monitored under UCMR-3, would a 2-year monitoring timeframe instead of the 3-year timeframe be likely to provide EPA sufficient information on seasonal fluctuations in contaminant occurrence? (See app. IV for full survey.)

[65] As noted earlier, EPA may have an opportunity to use partial UCMR3 data to inform its CCL3 regulatory determinations, but not enough UCMR3 data will be available to serve as a primary source.

[66] Eighteen of the contaminants on CCL3 were carried over from CCL1 and CCL2.

[67] EPA can make regulatory determinations outside of the 5-year cycle but has done so only once. In 2011, EPA made an out of cycle determination to regulate perchlorate.

[68] Environmental Protection Agency, "Regulatory Determinations for the Third Drinking Water Contaminant Candidate List, Presentation for Stakeholder Meeting" (Washington, D.C.: June 16, 2011).

[69] EPA first listed these contaminants on CCL1 and carried them over to CCL2 and CCL3. They are (1) acetochlor, (2) alachlor ESA, (3) disulfoton, (4) diuron, (5) MTBE, (6) metolachlor, (7) molinate, (8) nitrobenzene, (9) RDX, (10) terbufos, and (11) vanadium.

[70] The other four contaminants are microbes and have not been monitored yet because EPA does not have analytical methods available for them. Methyl bromide is listed by the name bromomethane in UCMR3.

[71] While EPA is authorized to make regulatory determinations outside of the 5-year cycle, in practice, it has done so only once.

[72] These contaminants are (1) disulfoton, (2) diuron, (3) MTBE, (4) molinate, (5) nitrobenzene, and (6) terbufos.

[73] We have previously reported that gaps in health effects data limit EPA's ability to make regulatory determinations and recommended that EPA develop a coordinated process for obtaining the necessary health effects and occurrence data. EPA's response to our recommendation did not acknowledge that any further steps are needed. See GAO-11-254.

[74] Environmental Protection Agency, "Drinking Water: Regulatory Determinations Regarding Contaminants on the Second Drinking Water Contaminant Candidate List," 73 Fed. Reg. 44,251 (July 30, 2008).

End Notes for Appendix III

[1] Drinking water types in the United States include groundwater sources, such as underground aquifers, and surface water sources, such as streams, rivers, and lakes.

[2] Volatile organic compounds, or VOCs, may occur naturally in the environment or occur only as a result of manmade activities; some VOCs have both origins. VOC contamination of drinking water supplies is a human-health concern because many are toxic and are known or suspected human carcinogens. EPA has issued drinking water regulations for eight VOCs.

In: Drinking Water and Unregulated Contaminants
Editor: Cynthia Romero
ISBN: 978-1-63321-164-3
© 2014 Nova Science Publishers, Inc.

Chapter 2

SAFE DRINKING WATER ACT: EPA SHOULD IMPROVE IMPLEMENTATION OF REQUIREMENTS ON WHETHER TO REGULATE ADDITIONAL CONTAMINANTS*

United States Government Accountability Office

WHY GAO DID THIS STUDY

Under 1996 amendments to the Safe Drinking Water Act, every 5 years the Environmental Protection Agency (EPA) is to determine for at least five contaminants, such as chemicals, whether regulation is warranted, considering those that present the greatest public health concern. Since 1996, EPA had not recommended any new contaminants for regulation until February 2011, when it reversed its controversial 2008 preliminary decision to not regulate perchlorate, an ingredient in rocket fuel and other products. GAO was asked to (1) evaluate the extent to which EPA's implementation of the 1996 amendments has helped assure the public of safe drinking water and (2) review the process and scientific analyses used to develop the 2008 preliminary regulatory determination on perchlorate. GAO analyzed relevant statutory provisions and regulatory determination documents and interviewed EPA officials.

WHAT GAO RECOMMENDS

GAO's 17 recommendations include that the EPA Administrator require (1) development of criteria to identify contaminants that pose the greatest health risk, (2) improvements in its unregulated contaminants testing program, and (3) development of policies or guidance to interpret the broad statutory criteria. EPA agreed with 2 recommendations but took the

---

* This is an edited, reformatted and augmented version of United States Government Accountability Office publication, No. GAO-11-254, dated May 2011.

position that developing guidance and taking the other recommended actions are not needed. GAO believes EPA needs to adopt all of the recommendations to better assure the public of safe drinking water.

WHAT GAO FOUND

Systemic limitations in EPA's implementation of requirements for determining whether additional drinking water contaminants warrant regulation have impeded the agency's progress in assuring the public of safe drinking water. EPA's selection of contaminants for regulatory determination in 2003 and 2008 was driven by data availability—not consideration of public health concern. EPA does not have criteria for identifying contaminants of greatest public health concern and based most of its final determinations to not regulate 20 contaminants on the rationale of little or no occurrence of the contaminants in public water systems. Moreover, EPA's testing program for unregulated contaminants—which can provide key data to inform regulatory determinations—has fallen short in both the number of contaminants tested and the utility of the data provided because of management decisions and program delays. In addition, EPA has not developed policies or guidance for interpreting the amendments' broad statutory criteria for selecting contaminants and making regulatory determinations, increasing the potential for inconsistent decision making. Also, the credibility of some of EPA's regulatory determinations is limited by a lack of transparency, clarity, and consistency of key documents. For example, EPA made decisions on nine contaminants relying on tests that were not sensitive enough to detect them at the agency's health risk benchmarks. Furthermore, EPA did not clearly and consistently disclose this limitation and its effect on EPA's analysis.

In making its preliminary regulatory determination on perchlorate in 2008, EPA used a process and scientific analyses that were atypical, lacked transparency, and limited the agency's independence in developing and communicating scientific findings. First, while an intra-agency workgroup typically makes recommendations to the Assistant Administrator for Water on whether to regulate evaluated contaminants, in this case, the Assistant Administrator directed the staff to develop a determination to not regulate and to support a specified exposure level as protective of all populations. This direction was outlined in an agreement between high-level officials at EPA and other federal agencies that is not part of the perchlorate regulatory determination record. Moreover, EPA adopted the National Academies' 2005 perchlorate health assessment—a foundation for EPA's regulatory determination—without using EPA's standard internal scientific review process. This assessment is controversial, especially its sufficiency to protect infants. Also, the credibility of EPA's exposure estimate for perchlorate, which is based on a novel analysis, is reduced by the lack of a comprehensive explanation of the methodology's limitations and uncertainties in the preliminary determination notice. Finally, according to key EPA scientists, the agency mischaracterized important scientific findings on the sensitivity of various age groups to perchlorate exposure. EPA scientists who managed the sensitivity analysis did not agree that it supported the conclusion that the selected exposure level was protective of all populations, which was one component of the aforementioned agreement between EPA and other federal agencies.

ABBREVIATIONS

ATSDR	Agency for Toxic Substances and Disease Registry
CDC	Centers for Disease Control and Prevention
DOD	Department of Defense
EPA	Environmental Protection Agency
HHS	Department of Health and Human Services
IRIS	Integrated Risk Information System
MTBE	methyl tertiary-butyl ether
NASA	National Aeronautics and Space Administration
NAWQA	National Water-Quality Assessment
NIRS	National Inorganic and Radionuclide Survey
OMB	Office of Management and Budget
PBPK	physiologically based pharmacokinetic
PCE	tetrachloroethylene
TCE	trichloroethylene
USGS	U.S. Geological Survey
VOC	volatile organic compounds

May 27, 2011

The Honorable Barbara Boxer
Chairman
Committee on Environment and Public Works
United States Senate

The Honorable Henry A. Waxman
Ranking Member
Committee on Energy and Commerce
House of Representatives

The Honorable Edward J. Markey
House of Representatives

The overall goal of the Safe Drinking Water Act, originally enacted by Congress in 1974, is to ensure that public drinking water is safe.[1] Nonetheless, more than 35 years later, the safety of drinking water remains a significant concern. For example, according to a 2010 Gallup survey on the environment, the safety of drinking water continues to be the environmental issue of greatest concern to Americans, with 50 percent worrying “a great deal” about drinking water pollution. In addition, Members of Congress continue to express concerns regarding the impacts of contaminated drinking water on public health, particularly on children.[2] While 89 contaminants have been regulated pursuant to the act, the number of potential drinking water contaminants is vast.[3] For example, as many as tens of thousands of chemicals may be used across the country, and the Environmental Protection Agency (EPA) has identified more than 6,000 chemicals that it considers the most likely sources of human or environmental exposure. The potential health effects of exposure to most of these chemicals, and the extent of their occurrence in drinking water, are unknown. Several studies since the 1980s have examined the occurrence of unregulated contaminants in public drinking water systems and their source waters. Typically small in scale—focusing on 60 or fewer contaminants—these studies have nonetheless detected hundreds of unregulated contaminants in public drinking water systems and source waters, some of which are known to have adverse health effects.

Under the Safe Drinking Water Act, EPA is authorized to regulate contaminants in public drinking water systems. Since 1974, EPA has implemented its drinking water program under three separate legislative frameworks—first under the initial statute and subsequently under major amendments in 1986 and 1996. Under the 1996 amendments, which remain in effect, EPA is to select for consideration those unregulated contaminants that present the greatest public health concern, evaluate their occurrence and the potential health risks associated with them, and decide whether a regulation is needed for at least five contaminants every 5 years. This regulatory determination process includes EPA's publication in the *Federal Register* of a preliminary decision on whether the agency will propose a drinking water regulation for each contaminant evaluated— called a preliminary regulatory determination—and provides for a public comment period, followed by a final decision, or regulatory determination, also published in the *Federal Register*.

In 2003 and 2008, EPA issued final regulatory determinations on a total of 20 contaminants, deciding in each case that a drinking water regulation was not warranted. Thus, EPA had not recommended any new contaminant for regulation under the provisions of the 1996 amendments until February 2011, when the agency issued a final determination to regulate perchlorate—reversing a separate preliminary determination to not regulate perchlorate that the agency had issued in 2008. EPA had first formally identified perchlorate, an ingredient in such products as rocket fuel and fireworks, as a contaminant that may require regulation in 1998. Perchlorate can interfere with the normal functioning of the thyroid gland by inhibiting the transport of iodide into the thyroid, an essential step in the synthesis of thyroid hormones. According to EPA's preliminary and final determination notices on perchlorate, iodide uptake inhibition from perchlorate exposure has been identified as a concern in connection with increasing the risk of neurodevelopmental impairment in fetuses of hypothyroid mothers.[4] In addition, poor iodide uptake and subsequent impairment of thyroid function in pregnant and lactating women have been linked to delayed development and decreased learning capability in their infants and children. As we recently reported, perchlorate has been found in water, soil, and sediment at varying levels in 45 states, as well as in the food supply.[5] Over the last decade, issues surrounding the health risks of and the potential drinking water regulation for perchlorate have generated considerable interest and debate among such federal agencies as the Department of Defense (DOD) and the National Aeronautics and Space Administration (NASA) that use perchlorate in carrying out aspects of their missions. Moreover, EPA's 2008 preliminary determination to not regulate perchlorate in drinking water was controversial, generating significant public comment. In August 2009, the EPA Administrator announced that the agency would consider additional public comments on alternative analyses regarding perchlorate prior to making its final regulatory determination. In February 2011, the EPA Administrator announced that the agency intends to propose a perchlorate drinking water regulation within 24 months of the regulatory determination—that is, by February 2013.

In this context, this report responds to your request that we conduct a review of EPA's implementation of the Safe Drinking Water Act's provisions on unregulated contaminants.[6] Our objectives were to (1) evaluate the extent to which EPA's implementation of the 1996 amendments' requirement for determining whether to regulate potentially harmful contaminants has helped assure the public of safe drinking water and (2) review the process and scientific analyses EPA used to develop its 2008 preliminary regulatory determination on perchlorate.

To assess EPA's implementation of the 1996 amendments' requirement to determine which potentially harmful drinking water contaminants should be regulated, we reviewed the statute, legislative history, and relevant amendments and analyzed relevant documentation, such as *Federal Register* notices and regulatory determination support documents;[7] EPA's information on the potential adverse health effects of, and the extent to which the public may be exposed to, individual contaminants in public drinking water systems; and public comments on EPA's determinations. We interviewed officials from EPA's Office of Water and Office of Research and Development. We also reviewed health effects and public drinking and source water occurrence information from other entities, such as the Department of Health and Human Services' (HHS) Agency for Toxic Substances and Disease Registry (ATSDR), California's Office of Environmental Health Hazard Assessment, and the U.S. Geological Survey (USGS). In evaluating EPA's implementation of the regulatory determination process, we analyzed information developed under two processes integral to the regulatory determination process—development of contaminant candidate lists and implementation of the unregulated contaminants monitoring rule—but did not evaluate EPA's implementation of them. To review the process and scientific analyses EPA used to develop a preliminary determination on perchlorate, we interviewed officials in EPA's Office of Water and Office of Research and Development and reviewed relevant *Federal Register* notices; public comments; and EPA documents, including documentation of the agency's review process for its Integrated Risk Information System (IRIS) assessment for perchlorate and its scientific analyses of (1) exposure to perchlorate in drinking water and related external peer review comments and (2) the sensitivity of various age groups to perchlorate exposure and related agency documents. Appendix I provides a more detailed description of our scope and methodology. We conducted this performance audit from March 2009 to May 2011, in accordance with generally accepted government auditing standards. Those standards require that we plan and perform the audit to obtain sufficient, appropriate evidence to provide a reasonable basis for our findings and conclusions based on our audit objectives. We believe that the evidence obtained provides a reasonable basis for our findings and conclusions based on our audit objectives.

BACKGROUND

The following provides information on the act, its requirements and amendments, completed regulatory determinations, EPA health advisories, and federal advisory committees for drinking water issues.

The Safe Drinking Water Act

Among other things, the Safe Drinking Water Act requires EPA to establish legally enforceable standards for public water systems—called national primary drinking water regulations—which generally limit the levels of specific contaminants in drinking water that can adversely affect public health. States typically have the lead role in implementing and enforcing these federal drinking water regulations.[8] Under state laws, some state environmental agencies have the authority to establish more stringent standards than federal regulations and regulate additional contaminants than federal regulations, while others lack these authorities.

EPA's regulatory actions under the Safe Drinking Water Act have varied over time as the agency's legal authority to determine which drinking water contaminants to regulate, if any, has changed from discretionary to prescriptive and back to largely discretionary. While the statute was enacted in 1974, EPA promulgated most of the existing national drinking water regulations under the framework established by the 1986 amendments, which mandated, among other things, that EPA regulate specific contaminants and established a testing program whereby EPA was to require public water systems to test for unregulated contaminants. As discussed later, with the 1996 amendments, Congress established a framework under which EPA is to periodically identify contaminants that may warrant regulation, and it revised EPA's authority related to the existing testing program for unregulated contaminants. The 1996 amendments mandate that EPA focus on unregulated contaminants that present the greatest public health concern. EPA completed its first cycle of regulatory determinations in 2003 and a second cycle in 2008, and plans to complete a third cycle in 2013.[9] As noted earlier, EPA had decided to not regulate any additional contaminants since the enactment of the 1996 amendments[10] until February 2011, when the EPA Administrator announced that the agency had made a final determination to regulate perchlorate. The Administrator also announced plans to develop a regulation addressing a group of carcinogenic volatile organic compounds (VOC)—chemicals such as industrial solvents. The announced plan to regulate carcinogenic VOCs stemmed from EPA's effort to revise the existing drinking water regulations for trichloroethylene (TCE) and tetrachloroethylene (PCE) and the Administrator's 2010 Drinking Water Strategy that includes a goal of addressing contaminants as groups rather than one at a time to enhance drinking water protection in a timely and cost-effective manner. According to EPA, the agency plans to include up to eight unregulated VOCs in this regulatory action covering a group of related contaminants.[11] (App. II provides additional information on EPA's regulatory actions under the statute as enacted in 1974, the 1986 amendments, and the 1996 amendments.)

In addition to requiring EPA to periodically make regulatory determinations on unregulated contaminants, the 1996 amendments also require that EPA identify and publish a list every 5 years of unregulated contaminants that may require regulation; the list is called the contaminant candidate list. EPA selects the contaminants for which it will make regulatory determinations from the relevant candidate list.[12] EPA has published three candidate lists of unregulated contaminants that may warrant regulation: in 1998, 2005, and 2009 (see table 1).[13]

Table 1. EPA's Contaminant Candidate Lists

Contaminant candidate list	Date published in the *Federal Register*	Number ofcontaminants on the list
1	March 2,1998	60
2	February 24, 2005	51
3	October 8, 2009	116

Source: Federal Register notices.

For each regulatory determination cycle (2003 and 2008), EPA developed a preliminary determination notice that provided its proposed determinations for the contaminants addressed in that cycle, as well as its rationale and primary support for individual determinations. The final determination notice for each cycle addressed public comments received on its preliminary determination notice and briefly summarized its final determinations. In the two completed cycles to date, EPA did not change any determination in response to public comments.[14]

EPA's regulatory determinations can have long-term implications for the safety of public drinking water. For example, a decision to not regulate tends to limit or remove the focus and resources that may be applied to further developing and evaluating data on the occurrence of the contaminant, as well as on treatment technologies to remove it from drinking water. On the other hand, a decision to regulate places emphasis on and commits resources: first, to development of a regulation and, later, to actions by some public water systems to limit public exposure to the contaminant. Specifically, if EPA determines that a regulation is needed for a contaminant, under the 1996 amendments, the agency has 24 months to publish a proposed regulation for comment and up to an additional 27 months to promulgate a final regulation. With regard to developing drinking water regulations, the 1996 amendments added a requirement that EPA conduct a cost-benefit analysis as part of the standard-setting process. The amendments further require that in carrying out the provisions concerning listing, selecting, and regulating contaminants, to the degree that an action is based on science, EPA use the best available peer-reviewed science and data collected by accepted or best available methods. EPA is also to ensure that the presentation of information on public health effects is comprehensive, informative, and understandable.

Once established, federal drinking water regulations generally apply to the approximately 153,500 public water systems that provide drinking water to at least 15 service connections or that regularly serve at least 25 people. Nearly 52,000 of these systems are community water systems that serve year-round residents, providing drinking water to approximately 294 million people. Small systems—those serving 3,300 or fewer people—account for most of the community water systems, while large systems—those serving more than 10,000 people—provide drinking water to 82 percent of the population served by community water systems. Similarly, more than three-quarters of community water systems have groundwater sources, such as underground aquifers, but a much larger percentage of the population receives water from systems that have surface water sources, such as streams, rivers, and lakes. Appendix III provides further information on public water system types, sizes, sources, and populations served.

EPA's Office of Water has primary responsibility for implementing the requirements of the Safe Drinking Water Act. Office of Water staff have described the agency's process for assessing contaminants being evaluated for regulatory determination as one that continues to evolve. While the regulatory determination process is not a rulemaking process, EPA uses the administrative process it established for developing and processing regulations to govern the development and processing of its regulatory determinations. Moreover, according to EPA, the Office of Management and Budget (OMB) views the regulatory determinations as equivalent to "significant" rulemakings under Executive Order 12866, which establishes procedures for OMB to review certain regulations.[15]

Statutory Criteria for Assessing Drinking Water Contaminants for Regulation

The 1996 amendments stipulate that EPA assess contaminants against statutory criteria to make determinations on whether they warrant regulation. In making regulatory determinations, EPA is to consider contaminants that present "the greatest public health concern," taking into account sensitive populations—such as children—that may be at greater risk of adverse health effects from exposure to contaminants in drinking water, among other factors. EPA's regulatory determinations are to be based on three broad statutory criteria, all of which must be met for EPA to decide that a regulation is needed:

- the contaminant may have an adverse effect on the health of persons;
- the contaminant is known to occur or there is a substantial likelihood that the contaminant will occur in public water systems with a frequency and at levels of public health concern; and
- in the sole judgment of the Administrator, regulation of such contaminant presents a meaningful opportunity for health risk reduction for persons served by public water systems.

Data on Occurrence and Health Risks of Unregulated Contaminants in Drinking Water to Support Regulatory Determinations

To assess unregulated contaminants against the statutory criteria, the Office of Water needs sufficient information on both (1) the occurrence of these contaminants in drinking water—called occurrence data—to assess the population potentially being exposed and the levels of that exposure and (2) the human health effects that may result from exposure to the contaminants in drinking water. Regarding occurrence data, the agency has stated that it needs occurrence data that can provide "a generally representative idea of known and/or likely occurrence in public water systems."[16] As required by the 1996 amendments, EPA has developed a new testing program for unregulated contaminants to collect nationally representative samples from a subset of public water systems on up to 30 contaminants every 5 years.[17] Thus, the 1996 amendments placed a limit on the number of unregulated contaminants that certain public water systems would be required to test for in each 5-year testing cycle.[18] In this report, we refer to the unregulated contaminant monitoring program as EPA's testing program for unregulated contaminants. EPA completed one testing cycle in

2005 and expects to complete the second cycle later in 2011. On March 3, 2011, EPA issued a proposed rule for the third cycle and plans to publish a final rule in 2012. The published testing program rules identify, among other things, the contaminants to be tested, the testing (analytic) methods to be used, and the time frames for the testing. As shown in table 2, data from the first testing cycle was available for EPA to use in developing its second cycle of regulatory determinations in 2008, and data from the second testing cycle will be available to support EPA's third cycle of determinations in 2013.

Table 2. Information on EPA's Unregulated Contaminants Testing Program

Unregulated contaminants testing cycle	Date testing program rules published in *Federal Register*	Date testing program cycle completed[a]	Related regulatory determination cycle and completion date
Cycle 1	September 1999	December 2005	Cycle 2, 2008
Cycle 2	January 2007	Fall 2011 (planned)	Cycle 3, 2013 (planned)

Source: GAO analysis of EPA data.

[a]We consider completion of a testing program cycle to be the date at which EPA presents the final results.

Regarding the information EPA needs to assess health risks from exposure to unregulated drinking water contaminants, the Office of Water generally considers agency-approved health risk assessments, such as those available from the agency's Integrated Risk Information System (IRIS), sufficient to characterize potential health risks for the purpose of making regulatory determinations.[19] IRIS assessments, which are critical to the drinking water program, provide EPA's toxicity assessments of contaminants that may cause cancer and those that may cause neurological or other noncancer effects, or both.

Health Reference Levels That EPA Develops to Assess Contaminants for Regulation

Using an IRIS or comparable toxicity assessment, the Office of Water calculates the *health reference level*—the estimated level of exposure to a contaminant in drinking water below which adverse health effects are not likely. The health reference level is critical to EPA's implementation of the 1996 amendments because it is the benchmark the agency uses to evaluate whether contaminants occur in public drinking water at levels of public health concern. That is, levels of contaminants in drinking water that exceed the health reference level are deemed to represent exposure that is of public health concern. The health reference level is determined in one of two ways depending on whether the potential health effects are cancer (i.e., carcinogenic) or other adverse effects (i.e., noncarcinogenic). For carcinogens, the Office of Water generally develops the health reference level using a quantitative estimate of carcinogenic risk—from, for example, an IRIS assessment—to calculate a concentration in drinking water equivalent to a one-in-a-million increased risk of getting cancer from a lifetime of exposure to a contaminant. In making this calculation, EPA applies its standard metrics for the weight of an adult—70 kilograms (154 pounds)—and the daily drinking water intake—2 liters (60 fluid ounces). For contaminants with noncarcinogenic adverse health effects, such as perchlorate and manganese, the Office of Water develops the health reference level using the IRIS *reference dose*, which is an estimate of the total daily oral exposure to a contaminant—for example, from food and water—that is not likely to cause "appreciable risk of deleterious effects during a lifetime" and is expressed as milligrams (or micrograms) per kilogram of bodyweight per day (mg/kg/day or μg/kg/day).20, [21] EPA generally uses the reference dose to calculate a daily safe dose of the contaminant in drinking water for a healthy adult by applying (1) its standard metrics for the weight of an adult and the daily drinking water intake and (2) an allocation of the estimated oral exposure to the contaminant from drinking water alone, called the *relative source contribution*. (Stemming from the general assumption that there is no safe level of exposure to a carcinogen, EPA has not used the relative source contribution in developing health reference levels for carcinogenic contaminants.)[22] Appendix IV provides examples of the calculations for health reference levels for carcinogenic and noncarcinogenic adverse health effects.

The Relative Source Contribution Component of the Health Reference Level

The Office of Water applies the relative source contribution estimate when calculating the health reference level for noncarcinogens to ensure that the level of a contaminant in drinking water, when combined with other sources of exposure (e.g. food and air) will not result in a total exposure for an individual that exceeds the reference dose. The relative source contribution has a significant impact on the health reference level that the agency derives for contaminants with noncancer adverse health effects. As shown in table 3, the lower the relative source contribution, the lower and more protective the health reference level. Conversely, a higher relative source contribution results in a higher and less protective health reference level.

Table 3. Comparison of the Impact on Health Reference Levels of Three Different Relative Source Contribution Factors, Assuming a Reference Dose of 0.5 Micrograms per Kilogram per Day

[(Reference dose (micrograms per kilogram per day)[a]	x	Body weight) (kilograms)	÷	Drinking water intake] (liters per day)	x	Relative source contribution (percentage)	=	Health reference level (parts per billion)[b]
[(0.5	x	70)	÷	2]	x	20	=	3.5
[(0.5	x	70)	÷	2]	x	50	=	8.8
[(0.5	x	70)	÷	2]	x	80	=	14

Source: GAO.

[a]1 microgram = 0.001 milligrams. Therefore, 0.5 micrograms per kilogram per day also equals 0.0005 milligrams per kilogram per day.

[b]The parts per billion unit is equivalent to the micrograms per liter unit. Therefore, 3.5 parts per billion is the same as 3.5 micrograms per liter.

According to EPA's guidance on methods for developing relative source contribution estimates,[23] in deciding what method to use, the agency must determine whether adequate data are available on the extent of exposures to the contaminant from drinking water and all other relevant sources, including other oral exposures, inhalation exposures, and dermal exposures. According to an Office of Water official, in most cases the agency lacks adequate data to fully characterize the extent of exposure from the various sources. In the absence of sufficient data, the guidance recommends the use of a conservative default assumption—that the relative source contribution from drinking water is 20 percent, leaving 80 percent of the total reference dose to account for all other sources of exposure. With the exception of perchlorate, EPA has used the default 20 percent when using a relative source contribution to determine the health reference level for its regulatory determinations.

If the agency determines it has adequate exposure data to estimate the relative source contribution, rather than applying the default assumption, according to the guidance, the agency may choose one of two methods— the percentage method or the subtraction method.[24] EPA characterizes the percentage method as conservative in that it preserves the proportion of existing daily exposure from drinking water relative to other exposure sources when allocating the reference dose. That is, if the daily exposure to a contaminant in drinking water is currently low compared to food, for example, the portion of the reference dose allocated to exposure from drinking water will also be relatively low. In contrast, the subtraction method allocates 100 percent of the reference dose among the various sources of exposure. Specifically, under the subtraction method, the daily exposure from all nondrinking water sources is first subtracted from the reference dose, and the remainder is allocated to drinking water, thereby generally resulting in a higher relative source contribution factor for drinking water than the percentage method. The guidance generally provides that, using either method, the relative source contribution selected should not be lower than 20 percent or higher than 80 percent. (Appendix V provides example calculations of the relative source contributions that result from each of these methods.)

EPA's Completed Regulatory Determinations

EPA completed two cycles of regulatory determinations in 2003 and 2008 addressing a total of 20 contaminants—in each case, it decided not to regulate. In addition, EPA addressed perchlorate in 2011—deciding to regulate using an out-of-cycle determination. First, in 2003, EPA made final regulatory determinations to not regulate 9 contaminants on its 1998 candidate list of 60 contaminants. Second, in 2008, EPA made final regulatory determinations to not regulate 11 of 51 contaminants on its 2005 candidate list and issued a preliminary determination to not regulate perchlorate. As discussed earlier, on February 11, 2011, EPA reversed its preliminary determination and made a determination to regulate perchlorate—representing the first determination to regulate a contaminant under the 1996 amendments. Table 4 identifies the contaminants for which EPA has made regulatory determinations to date.

Table 4. Drinking Water Contaminants with Final EPA Regulatory Determinations

Regulatory determination cycle	Contaminants
Cycle 1 (2003)	Acanthamoeba, aldrin, dieldrin, hexachlorobutadiene, manganese, metribuzin,naphthalene, sodium, and sulfate.

Table 4. (Continued)

Regulatory determination cycle	Contaminants
Cycle 2 (2008)	Boron; dacthal (DCPA) mono-acid degradate; dacthal (DCPA) di-acid degradate; DDE; 1,3-dichloropropene; 2,4-dinitrotoluene; 2,6-dinitrotoluene; EPTC (s-ethyl-dipropythiocarbamate); fonofos; terbacil; and 1,1,2,2-tetrachloroethane.
Out of cycle[a] (2011)	Perchlorate.

Sources: 68 *Fed. Reg.* 42898 (July 18, 2003); 73 *Fed. Reg.* 44251 (July 30, 2008); 76 *Fed. Reg.* 7762 (Feb. 11, 2011).

[a]The 2011 regulatory determination on perchlorate represents the first time EPA has made an out-of-cycle determination, issuing it while EPA is working on the third cycle of regulatory determinations expected to be completed in 2013.

EPA's Health Advisories

EPA's Office of Water has issued health advisories for different durations of exposure for regulated and unregulated drinking water contaminants, including some for which EPA made regulatory determinations to not regulate.[25] According to EPA documents, health advisories provide technical guidance on health effects, analytical methodologies, and treatment technologies to assist EPA regional offices, state governments, and public health officials in cases of emergency spills or contamination situations. EPA has issued new or updated health advisories for 9 of the 20 contaminants it determined did not warrant regulation under the 1996 amendments.[26] These advisories establish concentrations of contaminants at which adverse health effects are not anticipated to occur over specific exposure durations (1 day, 10 days, several years, and a lifetime). Drinking water health advisories are not legally enforceable standards and, according to EPA, are subject to change as new information becomes available. A key difference between the utility of health advisories for regulated contaminants and health advisories for unregulated contaminants is that states, localities, and consumers may obtain testing data on the levels of regulated contaminants from their public water systems, but information on levels of unregulated contaminants may be outdated or unavailable from public water systems because these systems are not typically required to test for the presence of unregulated contaminants.

EPA's Federal Advisory Committees for Drinking Water Issues

The Office of Water has two key EPA federal advisory committees that can assist it in implementing the numerous requirements of the Safe Drinking Water Act. First, EPA's Science Advisory Board's Drinking Water Committee, composed of non-EPA technical experts, provides independent advice and recommendations to EPA on the technical aspects of its drinking water program.[27] For example, in 2009, the Drinking Water Committee issued a report on EPA's processes for developing its third contaminant candidate list.[28]

In addition, EPA's National Drinking Water Advisory Council, a federal advisory committee created in 1974 by the Safe Drinking Water Act, also provides independent advice and recommendations to EPA on various aspects of its drinking water program. The council is composed of 15 members representing (1) state and local agencies concerned with safe drinking water, (2) water-related or other organizations and interest groups having an active interest in safe drinking water, and (3) the general public. In June 2000, the council provided EPA with recommended protocols for evaluating contaminants for regulatory determination, including a semiquantitative evaluation tool that highlighted the relative importance of various factors in making regulatory determinations.[29]

Systemic Limitations in EPA's Implementation of Requirements for Determining Whether to Regulate Additional Contaminants Have Impeded Progress in Helping Assure the Public of Safe Drinking Water

Since the enactment of the 1996 amendments to the Safe Drinking Water Act, EPA has made limited progress in prioritizing drinking water contaminants on the basis of greatest public health concern, and the lack of data on the public's exposure to potentially harmful drinking water contaminants and their health effects continues to limit EPA's ability to make regulatory determinations.

In addition, during the nearly 15 years since the 1996 amendments were passed, EPA has not developed policies or guidance providing its interpretation of, or guiding personnel in how to implement, the broad statutory criteria for selecting contaminants and making regulatory determinations on them. Moreover, the credibility of some of EPA's regulatory determinations is reduced because of a lack of transparency, clarity, and consistency in the regulatory determination notices and primary support documents.

EPA Has Neither Identified the Drinking Water Contaminants of Greatest Public Health Concern Nor Fully Used Its Authority to Obtain Data for Making Regulatory Determinations

The 1996 amendments require EPA to consider for regulatory determinations contaminants that present the greatest public health concern, but the agency has not effectively implemented this requirement. In addition, while EPA has made some progress in developing the occurrence and health effects data it needs, for many contaminants EPA lacks sufficient occurrence and health effects data to support regulatory determinations, which continues to limit its ability to make these decisions. Further, some management decisions and implementation delays have limited the extent and utility of the occurrence data EPA collected under its unregulated contaminants testing program. Moreover, as a result of the IRIS program's inability to provide timely health assessment data, EPA regulatory determinations have been delayed on some unregulated contaminants.

EPA Has Not Effectively Implemented the Act's Requirement to Prioritize Its Regulatory Determinations by Selecting for Consideration Contaminants of Greatest Public Health Concern

In light of the potentially large number of contaminants in public drinking water and limited government resources to address them, the 1996 amendments to the Safe Drinking Water Act directed EPA to consider for regulatory determination contaminants that present the greatest public health concern. As discussed earlier, the occurrence in drinking water and potential health effects of tens of thousands of chemicals that may be in use in the United States are largely unknown. Although the candidate list represents one level of prioritization by having EPA identify contaminants that warrant consideration for regulation from a larger universe, EPA officials told us that the Office of Water has not (1) further ranked or otherwise prioritized the contaminants on the list on the basis of public health concern or (2) prioritized contaminants on the basis of public health concern when selecting them for regulatory determinations. In fact, for 16 of the 20 regulatory determinations made through January 2011, including all 11 decisions for the second cycle of determinations completed in 2008, EPA based its decisions not to regulate on the rationale of no or limited occurrence—that is, EPA assessed public exposure to these drinking water contaminants as minimal. An EPA official described these determinations as addressing the "low hanging fruit"—rather than the contaminants of greatest public health concern. In making regulatory determinations, EPA has selected contaminants from its candidate lists for which it decided that sufficient occurrence and health effects data were available.[30] Consequently, data availability—not consideration of greatest public health concern—has been the primary driver of EPA's selection of contaminants for regulatory determinations. EPA officials told us that the agency has not needed to prioritize among contaminants because a lack of occurrence and health effects data, rather than agency resources, has limited the number of determinations. According to EPA officials and regulatory determination documents, most contaminants on the candidate lists have lacked sufficient occurrence data, health effects information, or both.

Lack of Coordinated Occurrence and Health Effects Data Continues to Limit EPA's Ability to Make Regulatory Determinations

EPA has acknowledged the significant gaps in its data for contaminants on its candidate lists. After EPA completed 20 regulatory determinations from 1996 through 2008, 40 of the 60 contaminants on its 1998 and 2005 candidate lists remained unaddressed as a result of insufficient data.[31] In 2009, EPA published its third and latest candidate list, which contains 116 contaminants[32]—including 18 from the previous lists.[33] EPA's *Federal Register* notice on the third candidate list indicated that the agency lacked sufficient occurrence or health effects data, or both, for making regulatory determinations for at least 100 of these contaminants. Further, in many cases, gathering sufficient data to address contaminants awaiting determinations has taken EPA more than 10 years, and obtaining data on other contaminants on the current list may well take decades.

Moreover, 17 contaminants that have been on all three candidate lists lack the needed occurrence data and health effects information. One such contaminant is RDX—a powerful explosive used by the U.S. military in thousands of munitions and classified by EPA as a possible human carcinogen. Although RDX first appeared on a candidate list in 1998, EPA still lacks the data it needs to make a regulatory determination.[34] Specifically, EPA expects to have nationally representative occurrence data on RDX from the second testing cycle by the fall of 2011, in time for the third cycle of regulatory determinations that EPA expects to complete in 2013. However, it is not known at this time whether the minimum reporting level of the test methods EPA is using to collect occurrence data on RDX under its testing program will be sufficiently sensitive to identify exposure at the health reference level, because the IRIS assessment on which the health reference level will be based is in progress. Regarding the IRIS assessment, we have previously reported that EPA started an IRIS assessment of RDX in 2000, suspended it to await research, and then restarted it in 2007.[35] Subsequently, EPA's Office of Research and Development placed the ongoing IRIS assessment of RDX and other contaminants on hold for about a year to enable the IRIS program to focus on completing what it defined as its priority assessments.[36, 37] In October 2010, an official from the Office of Research and Development told us that the IRIS program had just begun to work on the IRIS assessment again and estimated its completion in the third or fourth quarter of 2012. Under EPA's IRIS assessment process, this approximately 2-year time frame reflects EPA's estimate for completion of a standard, noncontroversial assessment.[38] However, RDX is among the contaminants EPA considers to have a greater level of controversy or visibility. As we reported in 2008, several IRIS assessments of contaminants of key concern to other federal agencies—stemming from potential impacts on agencies' operations and associated environmental cleanup costs—involved extensive interagency reviews and ultimately long delays.[39] Because RDX is such a contaminant of concern, the IRIS assessment may similarly be subject to extended reviews and delays.[40] Moreover, the potential for the continuation of previous IRIS assessment delays for this contaminant raises concerns. Thus, the outlook for a regulatory determination on RDX in 2013 is uncertain.

Other contaminants that have been on all three candidate lists that lack the needed occurrence and health effects data include terbufos, a pesticide that is highly toxic; and methyl tertiary-butyl ether (MTBE), a gasoline additive whose potential health effects EPA is currently assessing under its IRIS program. EPA has a risk assessment for terbufos but does not yet have nationally representative occurrence data on it.[41] The agency has nationally representative occurrence data for MTBE but lacks an IRIS assessment. We note that addressing data gaps for regulatory determinations generally requires coordinated and timely efforts between the agency's testing program and its IRIS program. However, while data on exposure and health effects from similar time frames are generally needed to develop credible determinations and drinking water regulations, the current process is not producing such data, largely because EPA is collecting nationally representative occurrence data on some contaminants to identify exposure at levels of public health concern but is not at the same time actively pursuing IRIS assessments or other credible sources for health effects information. For example, in the case of RDX, while the Office of Water is obtaining testing data on this contaminant, we found that the Office of Water did not list RDX among the contaminants for which it needed an IRIS assessment in responding to the Office of Research and Development's 2010 survey of program offices on IRIS assessments needs for the next few years.

Some Management Decisions and Implementation Delays Have Limited the Extent and Utility of the Occurrence Data EPA Collected under Its Unregulated Contaminants Testing Program

Although occurrence data on unregulated contaminants are critical for informed regulatory determinations and may also help EPA identify contaminants of greatest public health concern, EPA has not fully used its authority to gather these important data. To obtain information on how frequently and in what locations unregulated contaminants occur in public water systems for EPA's assessments of public exposure to such contaminants, the agency requires a subset of public water systems to test for the presence of selected contaminants either two or four times during a consecutive 12-month period during a testing cycle.[42] While the 1996 amendments authorized EPA to require certain public water systems to conduct such testing for the presence of up to 30 unregulated contaminants every 5 years, EPA's implementation of this authority has limited the agency's progress in obtaining this important information. Specifically, by the fall of 2011, EPA expects to have occurrence data for 51 contaminants the agency included in its first two testing cycles; however, according to the testing provisions in the 1996 amendments, EPA could have (1) obtained occurrence data for up to 60 unregulated contaminants and (2) begun the process of obtaining such data for up to 30 more.[43] The difference between the occurrence data that EPA could be obtaining under its authority to support its regulatory determinations and the data it currently has or is in the process of obtaining largely stems from the agency not fully using its testing authority, not adhering to the testing program schedule, and requiring testing for some contaminants that are not on its candidate list.

First, despite having the authority to require testing for up to 30 drinking water contaminants in each 5-year cycle, in implementing the first two cycles of the testing program, EPA required that only 51 contaminants be tested—thereby not availing itself of its authority to obtain occurrence data for 9 additional contaminants. EPA officials who manage the testing program said that EPA chose to limit the number of contaminants tested in the first two testing cycles so that, if additional emerging contaminants of concern were identified, the agency would be able to incorporate them into its testing schedule without exceeding the statutory limit of 30 per 5- year cycle. However, these officials also said they have now concluded that this approach was not practical or necessary, given the logistical and operational challenges the agency and water utilities would face in testing for additional contaminants and because the 1996 amendments provided other authority to test for other emerging contaminants, if needed. In August 2010, these officials told us they were likely to recommend that the agency use its full testing authority in the next testing cycle, and EPA did so—proposing to require testing for 30 unregulated contaminants in its March 3, 2011, proposed rule for the third testing cycle.

Second, delays in implementing the testing program have reduced its productivity in obtaining occurrence data on contaminants on the candidate list. If EPA had met the statutory time frames for the testing program,[44] it would likely have completed two testing cycles and begun the third cycle in 2009 by identifying up to 30 contaminants for which it would require testing to be conducted. Based on the first two rules, the testing would likely have started in January 2011. However, as of March 2011, EPA had obtained testing data on 26 contaminants from the first testing cycle and was still collecting final testing data for the 25 contaminants from the second cycle.[45] While the reasons for the delays in the testing program are outside the scope of this report, EPA officials who manage the testing program told us that the delays stemmed from implementing the 1996 amendments' new and technically challenging requirements for testing unregulated contaminants, including developing testing methods for selected contaminants, developing quality assurance and quality control standards for the tests, and certifying laboratories to conduct the tests. They indicated, however, that they should be in a position now to manage the program more efficiently, beginning with the planned publication in 2012 of the final list of contaminants to be tested in the third cycle and adhering to 5-year cycles in the future.

Third, EPA has required a sample of public water systems to test for some contaminants that are not on its candidate lists, which has further limited the capacity of the testing program to support EPA's regulatory determinations. According to EPA, the purpose of the testing program is to provide the data necessary to determine if a contaminant occurs with a frequency and at levels of public health concern to warrant regulatory determination; the testing program also informs the development of candidate lists, EPA's primary mechanism for identification of contaminants that may require regulation.[46] In the second testing cycle, 9 of the 25 contaminants selected for testing did not appear on the candidate list. As a result, EPA has required a subset of public water systems to test for 9 contaminants not on the list that represents the primary source from which EPA selects contaminants for regulatory determinations.[47] To the extent that EPA has tested for contaminants not on its candidate lists, the agency has reduced the capacity of its testing program to address the occurrence data gaps that EPA officials said have precluded the agency from making regulatory determinations for contaminants on the candidate lists.

EPA Used a Limited Testing Approach for More Than Half of the Contaminants Tested

In addition to decisions and program delays that have limited the total number of contaminants tested, EPA's decisions on the type of sampling approach to use have limited the utility of some of the testing program data obtained. Regarding sampling, EPA has used two approaches— "assessment monitoring" and "screening survey." EPA officials described assessment monitoring as "the gold standard" for obtaining sufficient data on national occurrence of drinking water contaminants to support regulatory determinations. This approach can provide nationally representative occurrence data with low levels of uncertainty.[48] Nonetheless, under the first two testing cycles, EPA required assessment monitoring testing—conducted at approximately 4,000 public water systems—for less than half (22) of the 51 contaminants in the unregulated contaminants testing program.[49] EPA has used the more limited "screening survey" for a majority of the contaminants tested to date—29—under the unregulated contaminants testing program.[50] The screening survey—which required testing at 300 and 1,200 public water systems for the first and second cycles, respectively—results in greater uncertainty because it tests a smaller sample of public water systems. In fact, the Office of Water did not make regulatory determinations for 12 contaminants tested under the screening survey in the first cycle of testing. In EPA's 2008 regulatory determination notice, the agency classified these 12 contaminants as having insufficient occurrence data for making regulatory determinations. Thus, nearly half of the contaminants in the first testing cycle were nonetheless found to have insufficient occurrence data for making regulatory determinations. EPA stated in a *Federal Register* notice that it increased the sample size of the screening survey in the second testing cycle so that "the data can be used to support regulatory determinations and rule development, if warranted."[51] For example, EPA testing program documentation states that if a contaminant is found with some significance during the screening survey, EPA may be able to forgo assessment monitoring and make a regulatory determination based on these data to protect public health more quickly. We note, however, that this screening survey will provide estimates with greater uncertainty than those provided by assessment monitoring. Despite EPA's intent that screening survey data indicating significant occurrence of a contaminant may be sufficient to support a regulatory determination, whether EPA finds such data to be sufficient will not be known until EPA announces whether it will proceed with determinations for the contaminants on the second testing cycle screening survey list—or why it will not proceed. This is an important point as three times as many contaminants on the current candidate list (12) will have second testing cycle screening survey data as will have the more robust assessment monitoring data (4) for EPA to consider as it develops the third cycle of regulatory determinations, expected to be finalized in 2013.

According to an EPA testing program document, assessment monitoring was selected for contaminants that could be tested for using analytical methods that "utilize widely available technologies," with decisions on which testing approach to use for each contaminant based primarily on the availability of the analytical methods used and the associated issue of laboratory capacity. The document also states that the screening survey "primarily targets contaminants with analytical methods that generally utilize more sophisticated technology that may not be widely established in drinking water laboratories."[52] However, in August 2010, EPA officials who manage the testing program said the number of laboratories with the capacity to conduct the more sophisticated testing technologies has grown, in part because recent concerns about low levels of pharmaceuticals found in drinking water increased demand for these more advanced technologies. In August 2010, the officials told us they now anticipate that most, if not all, of the testing under the next cycle will be conducted using assessment monitoring. EPA's March 2011 proposed rule for the third testing cycle proposes assessment monitoring testing for 28 of the 30 contaminants.[53]

EPA's Testing Methods Were Not Sufficiently Sensitive in Some Cases to Identify the Presence of Contaminants at EPA's Health Reference Level

In some cases, the benefit of the occurrence data collected under the testing program was reduced because EPA required public water systems to use testing (analytic) methods that were not sufficiently sensitive to identify the presence of contaminants at EPA's health reference level—the health benchmark that EPA uses in assessing whether to regulate specific contaminants. For 9 of the 20 contaminants for which EPA made regulatory determinations in 2003 and 2008, the minimum reporting level—the lowest level of a contaminant at which detections can be reported under testing protocols—exceeded EPA's health reference level.[54] As a result, occurrence of these 9 contaminants at a level higher than the health reference level but lower than the minimum reporting level may not have been detected.[55] For four of the nine contaminants, EPA obtained occurrence data in the first cycle of its testing program.[56] According to EPA officials, the minimum reporting levels were developed for these contaminants before the health reference levels were developed. These officials told us that the agency used its best professional judgment and the information available at the time in developing the minimum reporting levels that, in retrospect, were higher than the health reference levels.[57] According to a 2010 USGS report that presents information from that agency's National Water-Quality Assessment Program—a long-term effort to assess the status and trends of national water quality conditions[58]—the reporting level for contaminant occurrence data should be below the human health benchmark, which is analogous to EPA's health reference level,[59] to ensure that the tests are adequate to detect concentrations relevant to human health. Further, USGS has reported that when the reporting level exceeds the health benchmark, a contaminant may be present at a concentration greater than the health benchmark but remain undetected, resulting in greater uncertainty in evaluating the contaminant concentration in the context of public health. We note that EPA's testing program obtains data using minimum reporting levels that are often higher than those used by USGS in its National Water-Quality Assessment Program—ranging from 2 to more than 600 times higher.[60] In addition to the potential for underestimating occurrence at the health reference level, using high minimum reporting levels also prevents EPA from obtaining information on lower levels of contamination—such as at one-half the health reference level, as recommended by EPA's National Drinking Water Advisory Council, or the 1/10th health reference level used by USGS. According to the USGS report, these lower levels can provide "early and conservative indications of contaminant concentrations that may at some time approach or exceed benchmarks." While EPA may have different factors to consider in selecting methods for use in its testing program than USGS, the USGS report shows that more sensitive methods for testing unregulated contaminants have been available.

The IRIS Program's Inability to Provide Timely Health Assessment Data Has Delayed Some Regulatory Determinations

The other key gap in data that are critical for informed regulatory determinations is a lack of health assessment data on drinking water contaminants. This gap has been largely caused by long-standing productivity problems in EPA's IRIS program, which is managed by the Office of Research and Development. EPA established the IRIS program in 1985 to develop EPA consensus opinions about the health effects that may result from chronic exposure to various substances found in the environment, thereby helping EPA program offices reduce inconsistency in toxicity assessments and, therefore, risk assessments. The IRIS database contains toxicity assessments providing health effects data for more than 540 chemicals. The Office of Water typically uses these assessments, along with exposure assessments, to characterize the public health risks of exposure to a chemical or contaminant in drinking water.[61] IRIS assessments are a cornerstone of scientifically sound EPA decisions, policies, and regulations under a variety of statutes and programs, such as the Safe Drinking Water Act, the Clean Air Act, and the Superfund program.[62] We reported in March 2008 that EPA has not been able to keep its existing chemical toxicity assessments current or to complete assessments of the most important chemicals of concern.[63] We subsequently added transforming EPA's processes for assessing and controlling toxic chemicals as a high-risk area in our January 2009 report on governmentwide high-risk areas requiring increased attention by executive agencies and Congress, and we published our biennial update report on the high-risk areas in February 2011.[64] In 2009, EPA issued a revised IRIS assessment process in response to our recommendations.[65]

The inability of EPA's Office of Research and Development to provide the Office of Water with new and updated assessments in a timely manner has impeded effective implementation of EPA's regulatory determinations. From 1998 through 2008, the Office of Water lacked current IRIS assessments or other sufficient health information for 24 chemical contaminants on its candidate lists, and the Office of Research and Development completed assessments for only 2 of the 24.[66, 67] Moreover, the Office of Water's needs for health effects information for contaminants on the current candidate list, which contains about twice as many contaminants as the prior lists, have roughly doubled. Specifically, when publishing the third candidate list in 2009, EPA identified health effects information gaps for 44 of the 104 chemicals on the list. Further, this number may be understated because it excludes 24 contaminants for which EPA does not have an IRIS assessment or for which an IRIS assessment is either in progress or has not yet begun. It is not clear why EPA has not also categorized these contaminants as having health effects information gaps, given the challenges that the IRIS program continues to face in timely completion of assessments and the related uncertainty as to when these ongoing and planned assessments may be completed. Further, most of the 44 contaminants with health effects information gaps that EPA did identify (1) are not on the IRIS agenda (i.e., assessments are neither under way nor planned) and (2) have not been identified by the Office of Water as priorities for IRIS assessments. For about half of the 44 contaminants with health effects information gaps, EPA has indicated that assessments containing sufficient toxicity information to initiate an IRIS toxicity assessment may be available from other government entities, such as the World Health Organization, ATSDR, and the California Environmental Protection Agency.[68]

While problems with the IRIS program are beyond the purview of the Office of Water, to some extent the Office of Water's failure to develop and communicate its assessment priorities to the Office of Research and Development during this period may have contributed to the Office of Water's inability to obtain sufficient health effects information. Specifically, the Office of Research and Development periodically requested that program offices identify their IRIS priorities and needs throughout the 2000s, and as we reported in 2008, some EPA program offices responded and specifically requested that IRIS program management focus its resources on expediting the completion of specific ongoing IRIS assessments important to their missions.[69] However, our review of program offices' responses to the Office of Research and Development's periodic requests for priorities showed that the Office of Water did not respond to the requests. Further, while Office of Water officials told us that they met with Office of Research and Development staff to request status updates of key IRIS contaminants "on a number of occasions," the officials confirmed they did not take the initiative to communicate the office's IRIS assessment needs in writing. We note, however, that the Office of Water did respond to the IRIS program's request for input on IRIS assessment priorities in 2010. Also, in November 2010, EPA Office of Water officials said that the office is developing health assessments for some contaminants internally in lieu of depending solely on the IRIS program, a situation that highlights the need for EPA to effectively implement its revised IRIS assessment process. That is, when EPA lacks an effective IRIS process, individual EPA program offices may respond by developing health risk assessments of contaminants, using varying assessment methods and potentially producing different risk estimates—a problem EPA hoped to mitigate by creating the IRIS program.

EPA Lacks Policies or Guidance on Applying the Broad Statutory Criteria for Selecting Contaminants for Regulatory Determinations and Making the Determinations

Although the Safe Drinking Water Act requires EPA to select contaminants for regulatory determinations that present the greatest public health concern, EPA has not developed policies or guidance defining the characteristics that would constitute a contaminant of greatest public health concern or established a process for identifying such contaminants. Moreover, the Office of Water has not developed guidance on when and how to conduct additional analyses focused on the effects of drinking water contaminants on children and other sensitive subpopulations. While the statutory criteria from the 1996 amendments that EPA is to apply in making regulatory determinations are generally broad and open to interpretation, EPA has not developed policies or guidance for applying them. Finally, EPA has not developed guidance on, or a process for, reconsidering final regulatory determinations to not regulate.

EPA Has Not Defined Characteristics of Contaminants of Greatest Public Health Concern for the Purpose of Selecting Them for Regulatory Determinations

EPA has not defined the characteristics of contaminants of greatest public health concern or developed a process for prioritizing the contaminants on its candidate list for regulatory determination on this basis. As a result, EPA lacks criteria and a process for identifying those contaminants on its candidate list that pose the greatest public health concern. Contaminants of greatest public health concern could be defined on the basis of various characteristics. For example, a contaminant that poses a significant health hazard, such as cancer or a serious adverse neurological effect, could be considered a contaminant of greatest health concern. Along these lines, a 1993 EPA report to Congress on the drinking water program discussed criteria the agency could use in selecting additional contaminants for regulation, including that "contaminants posing acute risks or very potent chronic health risks would be more likely to be regulated even if occurrence tended to be low."[70] In addition, the occurrence or likely occurrence of a contaminant in many public drinking water systems or in source water could also characterize a contaminant of greatest public health concern. For example, an extensive USGS report[71] on the quality of source water from public-supply wells based on tests of water samples for 215 regulated and unregulated contaminants conducted from 1993 through 2007 identified 10 contaminants that occurred most frequently above USGS health benchmarks[72]—5 of the 10 were unregulated contaminants.[73] These USGS data are limited to the source water for public water systems served by groundwater, but they nonetheless represent the type of data available for EPA to consider in identifying contaminants on its candidate lists as those of greatest public health concern.[74]

Moreover, EPA's Science Advisory Board has recommended that EPA prioritize among the contaminants on the candidate list. Specifically, in a 2009 report on the agency's draft third candidate list, EPA's Science Advisory Board said that the draft list of 104 contaminants was too large to effectively prioritize for the purpose of making regulatory determinations. The Science Advisory Board recommended that EPA prioritize the contaminants on the list to help the agency meet its goal of selecting contaminants that "have the greatest opportunity to improve the safety of drinking water and protect public health." In its 2009 *Federal Register* notice publishing the final candidate list, EPA stated that it would continue to work to prioritize the 116 contaminants on the third candidate list, both for regulatory determination and for additional research and data collection. As noted earlier, it is important that such efforts relating to IRIS assessments and the unregulated contaminants testing program be effectively coordinated—that is, focused on priority contaminants of greatest public health concern.

The Office of Water Has Not Developed Guidance on Considering Adverse Health Effects of Drinking Water Contaminants on Children and Other Sensitive Subpopulations to Identify Contaminants of Greatest Concern or in Making Determinations

In selecting contaminants that present the greatest public health concern, under the Safe Drinking Water Act, EPA is to consider the effect of these contaminants on subpopulations at greater risk of adverse health effects, such as children. Further, Office of Water officials stated they are required to consider effects on sensitive subpopulations in making regulatory determinations. Such subpopulations, which may be at greater risk for adverse health effects from exposure to drinking water contaminants, may include infants, children, those with kidney or liver diseases or weakened immune systems, and the elderly. In fact, some of these subpopulations, such as children and individuals with liver or kidney disease, are identified as sensitive subpopulations for many of the contaminants EPA determined did not warrant regulation. For example, according to EPA documents, because children consume more water per unit of body weight than adults, they may be more highly exposed to toxic substances in drinking water and therefore at greater risk of adverse health effects than adults. In addition, children may have increased susceptibility following exposure to drinking water contaminants because they continue to develop both behaviorally and physiologically throughout childhood.

In 1995, EPA published its *Policy on Evaluating Health Risks to Children*, which states that the agency will "consider the risks to infants and children consistently and explicitly as a part of risk assessments generated during its decision making process," and to "the degree permitted by available data in each case, the Agency will develop a separate assessment of risks to infants and children or state clearly why this is not done."[75] In 2006, EPA developed a general guidance document for all EPA program offices on implementing its 1995 children's health policy,[76] as well as several technical guidance documents (in 2005, 2006, and 2008) that could help the Office of Water develop its own guidance specific to assessing the sensitivity of children to drinking water contaminants.[77] For example, EPA's 2005 guidance on assessing childhood exposures to environmental contaminants in general recommends that chronic risks be assessed by summing time-weighted exposures that occur at each life stage and states that adjustments for variations in toxicity may also need to be made for different age groups.78, [79] Finally, EPA has stated that in making a regulatory determination, "the [act] requires EPA to take into consideration the effect contaminants have on subgroups that comprise a meaningful portion of the general population (such as infants, children, pregnant women, the elderly, individuals with a history of serious illness or other subpopulations) that are identifiable as being at greater risk of adverse health effects than the general population."

Notwithstanding the requirements of the Safe Drinking Water Act and EPA's 1995 children's health policy, in developing the 2008 regulatory determinations, the Office of Water did not implement a specific approach to considering children's health. In addition, the Office of Water has not developed guidance for when and how to analyze the effects of drinking water contaminants on children—or other sensitive subpopulations—for the purposes of identifying the drinking water contaminants of greatest concern on which to make regulatory determinations and to ensure it consistently and explicitly considers risks to children in making these determinations. While EPA identified children as a sensitive population in 11 of the 20 regulatory determinations it completed in 2003 and 2008, Office of Water officials confirmed that for these 20 determinations, EPA did not develop separate health reference levels for children or make adjustments to its health assessments. Office of Water officials said, however, that they believe their evaluation of the health risks of contaminants takes into account sensitive subpopulations, including children, as required by the Safe Drinking Water Act. For example, the officials said that if they had determined children were "particularly sensitive" to the adverse health effects of contaminants being considered for regulatory determinations, they could have developed a separate assessment of risks for that population, using weight and drinking water intake for children, rather than the agency's standard approach using an adult's weight and water intake.[80] Office of Water officials said that, alternatively, they could have adjusted the reference dose to account for increased sensitivity. As discussed later, EPA's regulatory determination documentation does not explain how or whether the agency determined that a separate assessment for children or an adjustment to the health reference level was not warranted. In light of the children's risk issues discussed earlier—their consumption of more water per unit of body weight and other susceptibilities that may occur during childhood—EPA's approach to considering sensitive subpopulations in regulatory determinations may not fully account for the risks to children of exposure to drinking water contaminants and does not align with current agency guidance.

EPA Has Not Developed Policies or Guidance for Applying the Broad Statutory Criteria for Making Regulatory Determinations

The 1996 amendments also provide three broad criteria for EPA to use in making regulatory determinations, all of which must be met for EPA to determine that regulation is warranted. Notably, two of the criteria are so broadly stated that they could potentially be interpreted so as to lead to regulating all of the contaminants on candidate lists, some of them, or none of them. Specifically, the second statutory criterion—that a contaminant is "known to occur or there is a substantial likelihood that the contaminant will occur in public water systems with a frequency and at levels of public health concern"—is susceptible to varying interpretations. For example, different people may reasonably have differing views on the frequency and levels of occurrence that represent a public health concern. The third criterion—that regulation of the contaminant presents "in the sole judgment of the Administrator . . . a meaningful opportunity for health risk reduction"—is expressly discretionary, and similarly open to differing interpretations. Importantly, the Office of Water has not developed policies or guidance to help EPA staff apply these broad criteria. Guidance that might help EPA staff apply the criteria transparently and consistently could, among other things, (1) define or set thresholds or parameters for assessing whether a contaminant occurs, or is substantially likely to occur, in public water systems with a frequency and at levels of public health concern and (2) provide factors or characteristics of situations that would present meaningful opportunities for health risk reduction. We note that such guidance could also serve as the basis for an internal review mechanism to help EPA ensure consistent implementation of the statutory criteria. Office of Water officials could not describe examples of what would meet the three criteria beyond stating that "there are no bright lines" and that they would "know it when we see it." Without clarifying guidance, EPA's regulatory determinations lack transparency, and EPA is at risk of making inconsistent determinations, undermining the program's credibility and the agency's ability to assure the public of safe drinking water.

We note that, in 2000, EPA's National Drinking Water Advisory Council provided EPA with a suggested approach for making regulatory determinations, including a tool to transparently and semiquantitatively evaluate relevant data on known or likely occurrences of contaminants in drinking water. Under this approach, data on both known occurrences and the substantial likelihood of occurrence would be identified and weighted on the basis of such factors as number of detections in excess of the health reference level as well as in excess of one-half the health reference level, geographic distribution of occurrence, trends in the production and use of the chemicals, and direct releases into surface waters.[81] While EPA has generally provided the types of occurrence data identified by the council in its regulatory determination documents, it has not adopted a quantitative or other systematic method that would transparently identify how and to what extent these data are used to support the agency's determinations.

While the Office of Water has not established regulations or guidance for applying the broad statutory criteria, EPA appears to apply an informal policy that contaminants warranting regulation should occur in public water systems on a "national" scale. That is, as illustrated in the following examples, EPA officials and regulatory determination documents indicate that occurrence is, at least to some extent, evaluated from a national perspective.

- Some EPA officials serving on regulatory determination workgroups told us that a contaminant must occur "nationally" to warrant a determination to regulate.
- Documents supporting EPA's first cycle of regulatory determinations state that the consideration of geographic distribution "is important because the agency is charged with developing national regulations, and it may not be appropriate to develop [national primary drinking water regulations] for regional or local contamination problems."
- In response to public comments about certain preliminary regulatory determinations, EPA stated that contaminants "were evaluated in terms of national significance" and that EPA considers "both the extent of national occurrence and the severity of health effects" when deciding if a regulation would provide a meaningful opportunity for health risk reduction.
- In addressing public comments from a state that had requested EPA reconsider its preliminary determination to not regulate two contaminants (2,4-dinitrotoluene and 2,6-dinitrotoluene), EPA stated that these chemicals did not occur "nationally in public drinking water systems at health levels of concern" and that if a contaminant appears to be "highly localized," it does not meet the statutory criterion.

In the cases of 2,4-dinitrotoluene and 2,6-dinitrotoluene, the state requesting reconsideration of the preliminary determination reported to EPA that it had found high levels of 2,4-dinitrotoluene in groundwater (in one case, 200,000 times greater than EPA's health reference level) in numerous locations in their state, including around ammunition and military sites.[82] EPA's *Federal Register* notice does not disclose that the state detections reported to the agency, as well as the one detection of 2,4- dinitrotoluene identified by EPA's unregulated contaminants testing program, involved levels of the contaminant significantly in excess of EPA's health reference level.[83] In issuing its final regulatory determination, EPA declined to change its decision and referred the state to its revised drinking water health advisory, issued in conjunction with the determination. EPA's health advisory for these contaminants presented some data on contamination at military sites in two states, stating that these data demonstrate the risk for contamination of groundwater at military bases where dinitrotoluene munitions have been used or stored. Guidance addressing such issues as what circumstances would meet an occurrence characterization of "highly localized" could help EPA reach consistent judgments over time in cases such as these. Further, while EPA's regulatory determination notice acknowledges that the agency was aware that these contaminants may be found at some military sites, the notice and support documents do not discuss how and whether this information was used to assess occurrence and gauge the likelihood of occurrence, which might include considering how many military and industrial sites may potentially be contaminated with 2,4- and 2,6-dinitrotoluene.[84] We note that EPA's occurrence testing methodology is not designed to identify localized but high contaminant occurrences in public water systems that stem from specific land uses that may be associated with likely contamination—such as military and industrial activities—or associated with Superfund sites with known contamination and releases of these contaminants reported to EPA under its Toxics Release Inventory.

Notably, the Safe Drinking Water Act does not require that contaminants be found in public water systems on a national basis for an Administrator to find a meaningful opportunity for health risk reduction. In fact, other parts of the statute provide for relief from monitoring and flexibilities for instances in which a contaminant occurs in certain areas but not in others. Moreover, there is nothing in the act's committee reports suggesting that a contaminant need occur nationally to support a decision to regulate. Without EPA guidance providing a definition or parameters, an informal "national occurrence" standard is open to shifting interpretations, potentially affecting the consistency and credibility of EPA's decision making. Importantly, to the extent EPA is informally applying an unspecified national occurrence requirement for contaminants to be evaluated as occurring "with a frequency and at levels of public health concern," EPA is implementing a critical policy and interpretation of the Safe Drinking Water Act that has neither been defined nor subjected to public review.

Further, aside from stating in its regulatory determination notices that applying the third statutory criterion involves evaluating a contaminant's potential health effects and the related occurrence and exposure estimates at the "health level of concern," EPA has not articulated guidelines or thresholds for how it is to assess whether regulating a specific contaminant would provide a meaningful opportunity for health risk reduction. The absence of guidelines on what scenario or scenarios might illustrate "a meaningful opportunity for health risk reduction" increases the potential for inconsistent decision making and reduces the decisions' transparency.

EPA Has Not Developed Policies or Guidance on When the Agency Would Reconsider a Regulatory Determination to Not Regulate and What Process EPA Would Use to Do So

Office of Water officials told us that EPA could subsequently reconsider a determination to not regulate a contaminant on the basis of new scientific data. Office of Water officials said, for example, that a new IRIS health assessment on a contaminant could lead to a re-evaluation of a prior regulatory determination to not regulate that contaminant. EPA has not, however, developed any guidance on the circumstances that would trigger such a re-evaluation or on the process the agency would use in conducting a re-evaluation. In fact, in at least one instance—1,1,2,2- tetrachloroethane—new or revised scientific data became available after EPA's determination to not regulate, but the agency has not announced whether it will reconsider the determination. Specifically, subsequent to its negative regulatory determination in 2008, EPA's IRIS program finalized an updated health assessment in 2010, reclassifying this contaminant from a "possible" to a "likely" human carcinogen and decreasing the concentration in drinking water equivalent to a one-in-a-million increased risk of getting cancer. The new assessment would have produced a more stringent health reference level than the one EPA used in assessing the health risks of exposure to 1,1,2,2-tetrachloroethane for the regulatory determination. In addition, this contaminant is a VOC related through either production of, or degradation to, several other VOCs—namely, trichloroethylene (TCE), tetrachloroethylene (PCE), and vinyl chloride—that are (1) also associated with cancer, (2) currently regulated by EPA, and (3) among those contaminants EPA announced in February 2011 would be included in a planned new drinking water regulation of a group of VOCs. As there is no policy or guidance, it is not clear whether EPA will re-evaluate 1,1,2,2-tetrachloroethane as a result of the updated IRIS assessment or the planned action on carcinogenic VOCs in drinking water. EPA cannot ensure consistency in its re-evaluations of completed regulatory determinations in the absence of guidance on the circumstances under which, and the process the agency will use, to reconsider determinations. For example, potential process choices may include reconsidering such contaminants as a part of the agency's current regulatory determination cycle, adding the contaminants to the existing or next candidate list, or in the case of a planned regulation for a group of chemicals, adding the contaminant to that regulatory action.

The Credibility of Some of EPA's Regulatory Determinations As Presented in *Federal Register* Notices and Support Documents Is Limited by a Lack of Transparency, Clarity, and Consistency

The Safe Drinking Water Act requires EPA to ensure that, in its regulatory determinations, among other things, the presentation of information on public health effects is comprehensive, informative, and understandable. In addition, to the extent that EPA's regulatory determination notices and key support documents are transparent, clear, and consistent regarding the occurrence and health effects data the agency relied on, the credibility of the determinations is enhanced. However, for the regulatory determinations that EPA has made to date, some of the notices and support documents lack these key qualities.

EPA's Presentation of Health Effects Information on Some Contaminants Lacked Clarity, Consistency, and Transparency

EPA discussed the health risks to infants of exposure to manganese in drinking water in an inconsistent and, at times, incomplete manner in its regulatory determination documents. For example, EPA's 2003 regulatory determination support document for manganese states unequivocally that there are "no data to indicate children are more sensitive to manganese than adults." However, EPA's 2003 health effects support document for manganese discusses studies that identify an association between exposure to manganese in drinking water and learning disabilities in children and concludes that additional studies are needed to investigate the possibility that children are more sensitive than adults. In addition, while EPA's regulatory determination support document for manganese notes that infants and newborns may be potentially susceptible to manganese toxicity, this key document does not disclose that newborns may be exposed to high levels of manganese from infant formula or that these high levels of manganese in formula can be magnified when it is reconstituted with manganese-contaminated water. In contrast, the health advisory EPA issued in conjunction with the regulatory determination notice on manganese identifies "concerns for differences in manganese content in human milk and formula and the possibility of higher absorption and lower excretion in young infants." At a minimum, EPA's varying statements on the health risk that exposure to manganese in drinking water may pose to infants and children do not comport with the act's requirement to present information on health effects in a comprehensive, informative, and understandable manner.

In addition, EPA's presentation of the health risks of exposure to boron in the documents supporting its 2008 regulatory determination is also inconsistent and lacking in clarity. For example, in its regulatory support document, the Office of Water identified the primary adverse effects identified from studies of animals after chronic exposure to low doses of boron as generally involving the testes and the developing fetus. In this document, EPA stated that animal studies identify the developing fetus as potentially sensitive to boron and concluded that boron concentrations greater than the health reference level "might" have an effect on prenatal development.[85] In contrast, the Office of Water's May 2008 *Drinking Water Health Advisory for Boron*—developed in conjunction with the regulatory determination and published just 2 months before the regulatory determination was published—states that there are "compelling lines of evidence to suggest that the testicular morphological effects" reported in studies of animals are applicable to children. In this document, EPA also concluded that exposure to boron between birth and puberty may result in adverse cellular effects that would "affect testicular function" and that "testicular toxicity in males is the most sensitive endpoint relevant to children." In addition, a third related document—EPA's *Summary Document from the Health Advisory for Boron and Compounds*— provides an important warning regarding infants' exposure to boron in drinking water that is not included in either EPA's drinking water advisory for boron or its regulatory determination support document. Specifically, the summary document states that water containing boron "at levels above the HA [health advisory]" should not be used to prepare food or formula for infants. EPA does not identify which of the exposure duration health advisories it is referring to in this warning.[86]

EPA's Regulatory Determination Notices Lack Transparency and Clarity Regarding the Limitations of Health Advisories Issued in Conjunction with 10 Decisions to Not Regulate

As discussed earlier, EPA has issued new or updated health advisories for 9 of the 10 contaminants it determined did not warrant regulation under the 1996 amendments but which it decided warranted the issuance of health advisories. In its regulatory determination notices, EPA identified the purpose of the advisories variously, such as providing "guidance to communities that might be exposed to elevated concentrations" of the contaminants and "information to any states with public water systems that may have" contaminants at levels above the health reference level.[87] However, EPA's discussions of these health advisories in the regulatory determination notices for the contaminants lack clarity and transparency regarding the limitations of the advisories.[88] Specifically, the regulatory determination notices do not acknowledge that when EPA determines regulation is not warranted but a health advisory is needed, it will generally be up to states, localities, and consumers to determine whether such contaminant levels are found in public water systems in their jurisdiction. Importantly, because public water systems are not typically required to test for the presence of unregulated contaminants, information on the levels of the contaminants in individual public water systems may be outdated or unavailable. Some regulatory determination notices do, however, state that EPA "encourages" those states with public water systems that have contaminants at concentrations greater than the health reference level "to evaluate site-specific protective measures and to consider whether state-level guidance (or some other type of action) is appropriate." EPA officials have noted that individual states can promulgate their own drinking water regulations, but some states have legal or other constraints on their ability to regulate contaminants that EPA does not. For example, while some states—such as California and Massachusetts—have issued drinking water standards for contaminants that EPA has not regulated, others are statutorily prohibited from, or otherwise constrained in, enacting more stringent regulations than EPA has promulgated or promulgating their own drinking water regulations for contaminants that EPA does not regulate.[89] In addition, it would be difficult for many people to determine how much of these unregulated contaminants are present in their drinking water, which would be required, for example, to heed EPA's warning in some cases to not use drinking water with contaminants in excess of certain levels to prepare infant food or formula. As EPA has acknowledged in some of its health advisory documents, individuals may have to have their water tested by a laboratory for the presence of the contaminants since EPA typically does not require water systems to test for them. Moreover, according to EPA officials, the agency releases its drinking water advisories by posting them on its Web site and does not issue public notification of them, such as a press release, which potentially limits awareness of the health advisories.

EPA Explanations of the Occurrence Data EPA Relied on to Assess Known and Likely Occurrence of Contaminants in Drinking Water Lack Transparency, Clarity, and Consistency

EPA's regulatory determination notices and support documents generally lacked transparency and clarity and, in some cases, consistency, regarding the occurrence data the agency used in supporting its determinations. For example, in its preliminary regulatory determination notices, the Office of Water provided occurrence data on the number of public water systems, and the populations served by them, with contaminant detections (1) greater than the health reference level and (2) greater than one-half of the health reference level.[90] However, in its notices and support documents, EPA provides other data potentially relevant to assessing known or likely occurrence of contaminants in drinking water, such as data on releases to the environment as reported in the Toxics Release Inventory[91]; contamination at Superfund (National Priorities List) sites; and other data such as those provided by states, USGS, and others. The regulatory determination documents lack transparency because they do not explain the extent to which, if any, EPA used these data in its assessment of the contaminants' known or likely occurrence in public water systems. In addition, we found that EPA does not consistently disclose information about the occurrence of the contaminants at Superfund sites. For example, the regulatory determination notices and support documents did not report that ATSDR had found a particular contaminant at more than 50 percent of Superfund sites, while EPA did report such information from ATSDR about some contaminants found less frequently.

In addition, while EPA's regulatory determination notices for the 20 determinations completed in 2003 and 2008 present actual detections from unweighted occurrence data from public water systems that tested for the contaminants—rather than extrapolated estimates of national occurrence—EPA includes extrapolated estimates of national occurrence in its support documentation. However, it is not clear whether and to what extent EPA used these national occurrence estimates in assessing occurrence at levels of public health concern. For example, EPA's explanation in the notices that it does not provide estimates of national occurrence because presenting the data on actual detections is "the most straightforward and accurate way" to present the occurrence data does not address or clarify how the extrapolated data are used, if at all. Additionally, focusing on the actual detections rather than national projections in its determination notices has the effect of downplaying the potential occurrence of the contaminants nationally. For example, for manganese, EPA highlighted in the regulatory determination notice that 3.2 percent of groundwater public water systems serving approximately 39,000 people had at least one detection above the level of public health concern. In its support document for manganese, EPA reported that, according to its national estimates based on these groundwater data alone, approximately 2.3 million people could be affected.

Regulatory Determination Documents Lack Transparency and Clarity Regarding How EPA Determined Its Health Reference Levels Were Protective of Children

As noted earlier, EPA identified children as a sensitive subpopulation for 11 of the 20 contaminants with final regulatory determinations.[92] However, the regulatory determination notices and support documents in these cases lack clarity regarding how EPA determined that the health reference levels used to assess public health risk were adequately protective of this sensitive subpopulation. For example, in its 2003 regulatory determination notice addressing seven contaminants for which children were identified as a sensitive subpopulation, EPA stated that the agency had not yet determined a protocol for making a regulatory determination for a chemical for which body weight and drinking water intake of infants or a particular childhood age group would be the basis of a regulatory action. As discussed earlier, health assessments based on adult weight and drinking water intake may not fully account for the risks to children of exposure to drinking water because they consume more water per unit of body weight and may have other susceptibilities, as well. However, in its 2003 determination notice, EPA did not explain the potential effect of not developing separate health reference levels for children (or not making adjustments to its health assessments to reflect increased sensitivity) on its ability to ensure that the health reference levels used in the regulatory determinations were protective of children. In contrast to its 2003 determination notice, EPA was silent on the issue of separate assessments for children in its 2008 notice that included four contaminants for which children were identified as a sensitive subpopulation. As discussed earlier, Office of Water officials told us they would have developed separate assessments for children if they had determined children were "particularly sensitive" to the adverse health effects of contaminants being considered for regulation. However, EPA did not explain in its regulatory determination notices or support documents the basis for its determinations that children were not particularly sensitive to the adverse health effects of the contaminants considered for regulation—even for those contaminants that EPA had determined children are sensitive subpopulations; EPA also did not explain how the sensitivity of children can be evaluated in the absence of a separate assessment based on the weight and drinking water intake of children.

EPA's Regulatory Determinations Lack Clarity Regarding Its Reliance on Outdated and Limited Occurrence Data to Support Some Determinations

EPA relied primarily on nationally representative but older data from its National Inorganic and Radionuclide Survey (NIRS) on systems served by groundwater to assess (1) manganese occurrence in public water systems in 2003 and (2) boron occurrence in 2008. Discussing the use of the NIRS data (from the 1980s) in its 2003 *Health Effects Support Document for Manganese,* EPA said "these estimates are based on very limited and outdated data. The possibility exists that the number of people served by groundwater with [manganese] levels above the HRL [health reference level] could be higher than these estimates; however the data are lacking at this time to develop a more timely assessment." Moreover, EPA acknowledged that it did not have national data on manganese or boron occurrence in public water systems that use surface water—systems that, according to EPA, serve about 70 percent of community water system customers. As discussed in appendix VI, which provides additional information on the regulatory determinations for manganese and boron, the supplementary data EPA provided on surface water systems in its support documents for these contaminants were limited and, in the case of manganese, presented in a manner that may have understated the occurrence. For example, EPA acknowledged that the industry-sponsored survey data on boron were not statistically representative. Also, the agency characterized drinking water occurrence data on manganese from five states as showing "substantial low-level manganese occurrence," but did not reconcile this assessment with the data it presented showing that the percentages of state populations served by public water systems with levels that exceed the health reference level as ranging from 2.4 percent to 27.2 percent.[93] EPA's regulatory determination documents did not explain why, in light of these data deficiencies, it (1) deemed the occurrence data it relied on as sufficient for purposes of making regulatory determinations on these contaminants and (2) decided not to obtain current, nationally representative data on both ground and surface water occurrence of these contaminants through its unregulated contaminants testing program.

Regulatory Determination Documents Lack Transparency and Clarity Regarding EPA's Reliance on Minimum Reporting Levels Greater Than Its Health Reference Levels

As discussed earlier, for nine of EPA's 20 regulatory determinations, EPA based its conclusions on occurrence data for which some or all of the minimum reporting levels were above the health reference levels.[94] However, EPA's 2003 and 2008 regulatory determination documents for these contaminants lack transparency and clarity regarding the agency's use of minimum reporting levels in these cases and its potential effect on EPA's occurrence analyses. As table 5 shows, in these nine cases, the minimum reporting levels were from 1.25 to 2,200 times greater than the health reference levels.

Table 5. Nine Contaminants Whose Minimum Reporting Levels Exceeded EPA's Health Reference Levels

Contaminant	Health reference level (micrograms per liter - µg/L)	Minimum reporting level or range (micrograms per liter - µg/L)	Extent to which minimum reporting level was greater than health reference level
Aldrin[a]	0.002	0.1 – 0.84[b]	50 to 420 times
DDE[c]	0.2	0.8	4 times
1,3-Dichloropropene[c]	0.4	0.5	1.25 times
Dieldrin[a]	0.002	0.02 – 4.4[b]	10 to 2,200 times
2,4-Dinitrotoluene[c]	0.05	2	40 times
2,6-Dinitrotoluene[c]	0.05	2	40 times
Hexachlorobutadiene[a]	0.9	0.05 – 10[d] 0.1 – 1.5[b]	Up to 11 times at upper end of range Up to 1.7 times at upper end of range
Sulfate[a]	500,000	1 – 800,000 [b]	Up to 1.6 times at upper end of range
1,1,2,2-Tetrachloroethane[c]	0.4	0.01 – 10[d] 0.1 – 2.5[b]	Up to 25 times at upper end of range Up to 6.25 times at upper end of range

Source: GAO analysis of EPA data.
[a]Contaminant from regulatory determination Cycle 1, 2003.
[b]Unregulated contaminant monitoring program round 2 data. EPA provided estimates of the wide range of minimum reporting levels reported by states.
[c]Contaminant from regulatory determination Cycle 2, 2008.
[d]Unregulated contaminant monitoring program round 1 data. EPA provided estimates of the wide range of minimum reporting levels reported by states.

The regulatory determination notices for four of the contaminants from the 2003 regulatory determination cycle with minimum reporting levels greater than the health reference levels did not disclose that the occurrence data EPA provided as a basis for its conclusions could be understated because of this detection issue. The understatements could occur in these cases because some or all of the tests supporting the data presented were not sufficiently sensitive to detect occurrence at the agency's level of health concern (i.e., the health reference level).[95] Specifically, the regulatory determination notices present the occurrence data for the contaminants addressed in this determination cycle in a table identified as containing occurrence data for the number of systems with detections greater than the health reference level and one-half the health reference level, as well as the population served by these systems. However, there is no disclosure in the notice, the table, or a table note that, for four of the contaminants, the detections reported represent detections at minimum reporting levels that are higher than the identified health reference levels. In some cases, the undisclosed minimum reporting levels were significantly greater than the health reference levels. For example, for dieldrin, the agency relied on testing data obtained using minimum reporting levels ranging from 10 to 2,200 times higher than EPA's health reference level. Using these data, dieldrin was detected in 0.06 percent of samples (0.09 percent of public water systems). Using more sensitive tests with minimum reporting levels near and below EPA's health reference level, USGS's subsequent testing of source water for drinking water wells detected dieldrin in 3.1 percent of public well samples.[96] Importantly, essentially all of the detections were at levels above the health reference level, yet below EPA's minimum reporting levels. Thus, dieldrin would not have been found in these groundwater well samples if USGS had used the minimum reporting levels EPA used for its regulatory determination.[97] Further, while related regulatory determination support documents for the four contaminants with minimum reporting levels greater than EPA's health reference levels in some cases disclose the fact that the data provided are detections above the minimum reporting level, EPA's explanations are unclear and incomplete. Specifically, the documents state that the detections above the health reference level represent detections above minimum reporting levels because (1) the estimated health reference levels are lower than the minimum reporting levels or (2) a simple meaningful summary statistic is not available to describe the various reported minimum reporting levels, and to avoid confusion, minimum reporting levels are not reported. Importantly, the documents do not identify the minimum reporting levels, explain the impact of this limitation on EPA's occurrence analyses and its reliability, or explain why EPA considered these reporting levels appropriate for assessing occurrence at levels of public health concern.

The other five contaminants with some or all minimum reporting levels greater than EPA's health reference level were addressed in the 2008 regulatory determination cycle. For these contaminants, all of which are classified as possible or likely carcinogens, EPA did disclose in its determination notices that the minimum reporting levels used in the testing program were greater than the health reference levels—and that it therefore evaluated occurrence and exposure to the contaminants at the minimum reporting levels.[98] However, EPA did not clearly explain the effect that this approach has on the reliability of its analysis or clearly explain why the minimum reporting levels were appropriate for assessing occurrence at levels of health concern. That is, EPA explained in the regulatory determination notices that, for each of these contaminants, the minimum reporting levels were "within the 10^{-4} to the 10^{-6} cancer risk range," without providing the specific risk level value associated with the minimum reporting level for each contaminant. This range represents an increased risk of cancer to 1 in 10,000 persons (the 10^{-4} risk level) to 1 in 1 million persons (the 10^{-6} risk level). Since the minimum reporting levels of the different contaminants fall at different points within the risk range, EPA's assessments of occurrence and public exposure to contaminants at levels of public health concern for carcinogens are based on differing risk levels. For two contaminants for which we calculated the specific risk level values, the minimum reporting levels would limit cancer risk to approximately 1 in 25,000 persons; in contrast, EPA generally establishes health reference levels to limit cancer risk to 1 in 1 million persons. EPA also is not clear in its determination documents about (1) its rationale for applying risk standards for some contaminants that are less stringent than for others to identify occurrence at levels of public health concern or (2) the effect of the differing standards on the reliability of its estimates of occurrence at levels of public health concern.

EPA Lacked Consistency and Clarity in Making Determinations When IRIS Assessments Were Either in Process or Needed to Be Updated

An EPA document listing 14 contaminants the agency was evaluating for regulatory determinations in 2008 indicated that there was a "strong possibility" that several of the contaminants may not make the cutoff date for regulatory determinations because of outstanding risk assessments. In fact, the agency decided not to consider MTBE or bromobenzene for regulatory determination in 2008 because risk assessments for these contaminants were in process. However, EPA did make a regulatory determination in 2008 on 1,1,2,2-tetrachloroethane despite an update to an existing IRIS assessment being underway, without explaining why the agency did not await its completion.[99] As discussed earlier, the IRIS assessment for 1,1,2,2-tetrachloroethane was finalized in 2010 and would have produced a more stringent health reference level than the one EPA used to assess the health risks of exposure to this contaminant in its 2008 regulatory determination.

THE PROCESS AND ANALYSES EPA RELIED ON TO SUPPORT ITS PRELIMINARY DETERMINATION ON PERCHLORATE WERE ATYPICAL, LACKED TRANSPARENCY, AND LIMITED THE AGENCY'S INDEPENDENCE IN DEVELOPING AND COMMUNICATING ITS SCIENTIFIC FINDINGS

An understanding of the process and analyses EPA used to develop its 2008 preliminary determination to not regulate perchlorate can provide context for EPA's 2011 decision to reverse its preliminary determination. In addition, key analyses EPA relied on for its preliminary determination to not regulate perchlorate remain relevant to EPA's subsequent determination to regulate perchlorate. We found that in making its preliminary determination on perchlorate, EPA used a separate process that was less inclusive, less transparent, and more directive than the process it used to develop its other regulatory determinations. In addition, the agency used nontraditional approaches to three key analyses it relied on to support its preliminary determination. First, in developing an IRIS assessment for perchlorate, EPA established a reference dose on the basis of recommendations from the National Academies[100, 101] but subjected it to a more limited review than the agency's standard IRIS assessment review process. Second, to determine the relative exposure to perchlorate from drinking water versus food, EPA relied on an exposure estimate that it developed using a novel analysis and then used a nontraditional method to calculate the relative source contribution. Third, according to key EPA scientists, the agency mischaracterized important scientific findings that emerged from the novel analysis it conducted to determine the sensitivity of various age groups, particularly infants and children, to the agency-calculated health reference level. Furthermore, EPA's independence in developing and communicating its scientific findings was limited by the agency's acceptance of external input on its preliminary determination notice for perchlorate.

EPA Used a Less Inclusive, Less Transparent, and More Directive Process in Developing Its Preliminary Regulatory Determination on Perchlorate Than Its Usual Process

Perchlorate was initially among the contaminants EPA was evaluating in its second cycle of regulatory determinations, which were to be finalized by 2008. However, in 2007, when EPA published its preliminary determinations on these contaminants, which the agency was reviewing under its usual process for developing regulatory determinations, EPA indicated that it was not making a preliminary determination on perchlorate at that time, stating that additional information may be needed to more fully characterize perchlorate exposure. EPA stated that it would continue to seek and evaluate the information it needed—primarily regarding exposure to the contaminant—and would issue a preliminary determination on perchlorate either in time for it to be finalized along with the other contaminants' determinations in 2008 or as soon as possible following final determinations for the other contaminants. In conjunction with EPA's decision to issue a separate preliminary regulatory determination for perchlorate, EPA officials decided that the agency's continuing deliberations would be conducted using a different process, which differed substantially from its usual process "because of the sensitive nature of the perchlorate regulatory determination."

In developing its regulatory determinations for the first two determination cycles (2003 and 2008), EPA used its usual process, which the Office of Water initiated by establishing intra-agency regulatory determination work groups.[102] The work groups were composed of professional staff with relevant expertise from various EPA program, support, and regional offices.[103] Under this process, the intra-agency work groups identified contaminants on the candidate list for which it determined sufficient information was available to characterize both the potential health effects and the known or likely occurrence of the contaminants in drinking water; analyzed the health effects information and occurrence data for each selected contaminant with respect to the three statutory criteria; presented regulatory determination options[104] for each contaminant to the Office of Water's Assistant Administrator and the Assistant and Regional Administrators from offices participating on the work group, who selected a final option for the regulatory determination; and developed draft notices for review and obtained written position statements, generally in the form of concurrence memorandums, from all Assistant and Regional Administrators from participating offices.[105]

Under this usual process, after obtaining concurrence within EPA on the draft regulatory determinations, the Office of Water sent draft notices of its preliminary determinations to OMB for review, which could solicit and coordinate input from other federal agencies. Upon addressing comments from OMB and other federal agencies, as applicable, and receiving final clearance from OMB and final approval from the EPA Administrator, EPA published the preliminary regulatory determinations in the *Federal Register*, provided a public comment period, and subsequently published the final regulatory determinations in the *Federal Register*.[106] Throughout this process, EPA documented the meetings and decisions of the intraagency work group, concurrence memos from office leadership, and communications between EPA and external reviewers.

In contrast to EPA's usual process, which is managed by a work group of professional staff with relevant expertise from across the agency, EPA officials decided that the agency's continuing deliberations on perchlorate would be managed by a less inclusive, small group of high-level officials, such as the Deputy Administrator and several Assistant Administrators.[107] This group of high-level officials managed the regulatory determination process for perchlorate both within EPA and externally with the Perchlorate Interagency Working Group,[108] whose work was coordinated by the Council on Environmental Quality.[109] According to an EPA briefing document, the Perchlorate Interagency Working Group was established in 2002 "to identify and help resolve perchlorate science and science policy issues."[110] EPA officials told us that one of the aims of the Perchlorate Interagency Working Group was a "no-surprises policy" concerning any issues related to perchlorate. Thus, in addition to reviewing EPA's activities and analyses related to the perchlorate preliminary regulatory determination, the Perchlorate Interagency Working Group reviewed multiple products such as articles for scientific journals prior to dissemination by researchers and scientists from EPA and other agencies, such as the Centers for Disease Control and Prevention (CDC) and Food and Drug Administration. In addition, while the usual process for regulatory determinations provides multiple opportunities for input from participating EPA offices and results in documentation of meetings and analyses, the activities and decisions of the small group of high-level officials managing perchlorate were less inclusive and transparent. That is, relevant EPA staff, such as Office of Water managers, scientists conducting perchlorate analyses requested by the Assistant Administrator of the Office of Research and Development and the Perchlorate Interagency Working Group, officials from the Office of Children's Health Protection, and regional office staff from areas with extensive perchlorate contamination, had limited input into the preliminary regulatory determination. Moreover, according to an EPA official, meetings of the small group of high-level officials were not documented, and while the usual process involved developing a document providing the supporting analysis for the determination for each contaminant, EPA did not prepare such a document for the perchlorate preliminary determination. Also, in contrast to EPA's usual process, according to EPA officials, the Office of Water provided updates on the status of the perchlorate regulatory determination to the leadership of the EPA program and regional offices not included in the small group—such as the Office of Prevention, Pesticides, and Toxic Substances and the 10 regional offices—rather than seeking input and agreement from them. Specifically, agency program and regional offices were not asked by the small group of high-level officials to provide concurrence on the draft preliminary perchlorate determination.

EPA further limited participation by relevant EPA staff. For example, EPA did not include the Office of Children's Health Protection in its small group of high-level officials despite EPA's and the National Academies' conclusion that iodide uptake inhibition from perchlorate exposure had been identified as a concern in connection with increasing the risk of neurodevelopmental impairment in fetuses of pregnant women with iodine deficiency and to developmental delays and decreased learning capability in infants and children.111, 112 The lack of participation by the Office of Children's Health Protection in developing the preliminary regulatory determination for perchlorate is noteworthy for two reasons: (1) This office was established in 1997 to work with EPA program and regional offices to promote a safe and healthy environment for children by ensuring that all regulations, standards, policies, and risk assessments take into account risks to children[113] and (2) the Safe Drinking Water Act directs EPA to consider at several key points in the regulatory determination process the effect of contaminants on such subpopulations as infants, children, pregnant women, and others that may be at greater risk of adverse health effects than the general population as a result of exposure to those contaminants, among other factors.[114] Overall, by excluding relevant EPA offices from a more participatory role, the agency did not avail itself of the expertise that resides in those offices.

Finally, in contrast to the usual process EPA used for its regulatory determinations, in which EPA staff with relevant expertise develop and submit options to the Assistant Administrator for the Office of Water for review and selection, the Assistant Administrator directed the Office of Water staff in developing the preliminary determination for perchlorate. Specifically, according to officials in the Office of Water and the Office of General Counsel, in August 2008 the Assistant Administrator directed the staff to draft a preliminary determination that reflected the agency's decision to not regulate perchlorate and to support it with a detailed and specific rationale that EPA and other members of the Perchlorate Interagency Working Group had agreed to, under the leadership and coordination of the Council on Environmental Quality.[115] EPA Office of Water officials told us that they believed this agreement—which is not part of the record for the preliminary regulatory determination—was developed by senior officials from the Council of Environmental Quality, the Department of Health and Human Services (HHS), EPA, OMB, and the U.S. Department of Agriculture. The agreement focused on how EPA should address the key science issues concerning perchlorate in its preliminary regulatory determination and specified (1) a health reference level of 15 parts per billion of perchlorate in drinking water and (2) the rationale for EPA to support the conclusion that this health reference level would be protective of pregnant women and their fetuses as well as of infants and children.

EPA Established a Reference Dose for Perchlorate on the Basis of the National Academies' Recommendations, but Subjected It to a More Limited Review Than the Agency's Standard IRIS Assessment Review Process

In developing an IRIS assessment of perchlorate, EPA established a reference dose on the basis of the National Academies' recommendations, but subjected it to a more limited review than the agency's standard IRIS assessment review process. Typically, to establish a reference dose for a contaminant, EPA conducts an IRIS assessment, a draft of which undergoes internal and external scientific peer reviews and is also made available for public comment in the *Federal Register* before being finalized. In its initial effort to develop an IRIS assessment of perchlorate, EPA followed its usual process, and in 2002, the agency provided a draft IRIS assessment to the public for comment.

EPA's 2002 draft IRIS assessment of perchlorate—which proposed a reference dose of 0.03 micrograms per kilogram of body weight per day and a related drinking water equivalent level of 1 part per billion[116]—drew significant attention, including from such federal agencies as DOD, the Department of Energy, and NASA, because of the implications such a level could have on their operations if EPA were to develop a drinking water regulation for perchlorate.[117] According to a senior EPA official, the controversy that arose over the draft IRIS assessment of perchlorate "was like nothing I had ever seen or have seen since." As a result of the divergent views between EPA and the other federal agencies, the Administrator of OMB's Office of Information and Regulatory Affairs urged the four interested agencies to convene a National Academies panel to review the draft IRIS assessment. Convened in October 2003, the panel conducted this review and issued its report in January 2005.[118]

In its 2005 report, the National Academies made several key recommendations to EPA on the basis of a different study from those on which EPA had based its draft IRIS assessment. The National Academies' recommended reference dose was more than 20 times higher than the one proposed in EPA's 2002 draft IRIS assessment and, according to an IRIS official, was a recommendation that EPA neither specifically requested nor expected the panel to issue.[119] However, the IRIS official also told us that in discussions regarding the National Academies' recommendations, EPA management indicated that quickly incorporating the National Academies' perchlorate recommendations into the IRIS assessment would help the agency bring some closure to the controversy surrounding its draft assessment. In February 2005, just

5 weeks after the National Academies issued its report, EPA adopted the National Academies' recommendations with unprecedented expediency, finalizing its IRIS assessment on perchlorate using the National Academies' recommended reference dose. Other EPA officials told us that this swift action was possible because the National Academies had "taken the unprecedented step of completing what amounted to a toxicological review itself and derived a recommended reference dose."

EPA's final internal review of the revised IRIS assessment for perchlorate—termed a consensus review—differed from the agency's usual consensus review process for IRIS assessments at that time in terms of scope, time frames, and public comment. The scope of the review was limited in that the IRIS program did not seek input from consensus reviewers on the *scientific basis* for the assessment as it typically does. Instead, the IRIS program director stated in the request for consensus reviews that because the assessment had undergone several external peer reviews, including a recent review by the National Academies, the purpose of the consensus review of the revised IRIS assessment for perchlorate was to "ensure that the science in the IRIS Summary is clearly summarized and not inconsistent with the major conclusions of the [National Academies'] report."[120] At least two EPA offices essentially opted out of the consensus review process because of this limitation, which was a significant departure from the usual IRIS consensus practice.[121] For example, the Office of Children's Health Protection's response to the IRIS program's request for consensus review stated that because the IRIS assessment did not represent the "traditional end-of-process consensus review for IRIS," that office chose to neither approve nor disapprove the IRIS assessment, opting instead to identify concerns about the consideration of children's health in the IRIS perchlorate assessment, as discussed later. A second difference between the consensus review for the draft IRIS assessment and the typical consensus review was the speed with which it was conducted. The IRIS program allowed reviewers only 1 week—rather than the 45 days that is typically afforded—to complete their reviews before finalizing the revised assessment and posting it on its Web site.[122] Finally, while EPA had provided a public comment period for its 2002 draft IRIS assessment before the National Academies review began, it did not offer a public comment period on the 2005 revised assessment, even though it was based on different data than the 2002 draft assessment. Specifically, EPA's 2002 draft assessment was based primarily on data from several toxicological (animal) studies, while the National Academies' 2005 assessment was based primarily on the results of a single 14-day clinical study from 2002—the Greer study, which was funded by the National Institutes of Health and perchlorate manufacturers and users.[123]

EPA has received both internal and external criticisms of its decision to accept the National Academies' recommended reference dose as its IRIS assessment. These criticisms—which generally focused on limitations of the 2002 Greer study from which the National Academies derived its reference dose and the extent to which EPA mitigated those limitations to help ensure that sensitive subpopulations were adequately protected— were communicated to EPA before it finalized its IRIS assessment and 3 years before the agency issued its preliminary determination for perchlorate. For example, in 2005, the Office of Children's Health Protection articulated concerns in a memorandum to the IRIS program director during the week allowed for consensus review for the IRIS assessment that the uncertainty factor[124] of 10 that the National Academies used in developing its proposed reference dose might not sufficiently protect the sensitive subpopulations, which the National Academies identified as including fetuses, preterm newborns, infants, and developing children. The National Academies concluded this uncertainty factor was sufficient because its recommended reference dose was based on a nonadverse effect that was a precursor to any adverse effects that may result from exposure to perchlorate. However, the Office of Children's Health Protection stated in its memorandum to the IRIS program that the Greer study—which examined the effect of perchlorate exposure on iodide uptake inhibition of the thyroid in 37 healthy adults—did not provide data to consider variability among humans at different life stages, during pregnancy, and as a result of health status (such as hypothyroidism)—or even to consider variations among healthy adults. (In fact, the National Academies' recommended reference dose was derived from data on a subgroup of study participants comprised of seven healthy, nonpregnant adults.[125]) As a result, the Office of Children's Health Protection expressed its concern that the uncertainty factor of 10 the National Academies' recommended to account for variations among humans may not be adequately protective of sensitive subpopulations. The Office of Children's Health Protection also stated concerns that the effects observed during the 2-week Greer study may not be indicative of potential effects from lifelong exposure to perchlorate. Consensus reviewers from two other EPA offices identified similar concerns. For example, a consensus reviewer from a regional office, while expressing an understanding that accepting the National Academies' reference dose reflected "the best course of action for the agency, given the history of this compound," nonetheless provided substantive comments on a number of scientific issues, including the uncertainty factor. This reviewer stated that the National Academies' uncertainty factor apparently reflects a conclusion that pregnant women, pregnant women with subclinical hypothyroidism, preterm infants, and newborns would be no more sensitive to perchlorate than nonpregnant adult women and men—a conclusion the reviewer stated was questionable. Nonetheless the IRIS Summary that EPA posted on its Web site shows that the agency adopted the National Academies' reasoning, including its recommended uncertainty factor, which did not address the concerns raised by the Office of Children's Health Protection or reviewers from other offices.

In addition, in September 2005—3 years prior to EPA's preliminary determination for perchlorate—public health officials from Connecticut and Maine published a detailed critique of the Greer study that expressed concerns similar to those raised by the Office of Children's Health Protection, as well as additional concerns.[126] For example, these public health officials stated that the National Academies' consideration of uncertainties was insufficient. They said, among other things, that in addition to the uncertainty factor of 10 that the National Academies applied to account for variability among humans, an additional 3- to 10- fold uncertainty factor was warranted to account for the potential for greater toxicity to breastfeeding newborns,[127] as well as the potential for greater perchlorate toxicity resulting from exposure lasting more than 2 weeks.[128] These state public health officials concluded that risk assessors should carefully evaluate whether the IRIS reference dose was the most appropriate value for assessing perchlorate risk.

The concerns regarding the reference dose in the IRIS assessment noted earlier continued to be raised when EPA issued its preliminary regulatory determination for perchlorate in 2008. This preliminary determination generated substantial public comment regarding EPA's assessment of the health risks to sensitive subpopulations—issues that had been raised when EPA first adopted the reference dose. For example, in official comments on EPA's preliminary regulatory determination on perchlorate, officials from the states of Washington and Oregon asked EPA to reconsider its decision, citing, among other things, their view that the 10-fold uncertainty factor EPA used to derive a reference dose and to formulate a health-based drinking water level was insufficient. As with other comments, the officials stated that a higher uncertainty factor was warranted given the data gaps in the limited studies available on perchlorate. The state officials noted that the state of Massachusetts's risk assessment of perchlorate used a higher uncertainty factor—100 versus EPA's 10. In fact, two states that promulgated perchlorate limits for drinking water and one state that proposed such limits each relied primarily on the Greer study but, because of variations among the states regarding the values they chose for three key variables,[129] developed more stringent limits than EPA's assessment. Specifically, EPA's health reference level for purposes of its preliminary regulatory determination was 15 parts per billion of perchlorate per liter of water, while California's health limit is 6 parts per billion, Massachusetts's is 2 parts per billion, and New Jersey's proposal was 5 parts per billion.[130] Nonetheless, in its public comment response document associated with EPA's February 2011 final regulatory determination for perchlorate, EPA stated that it "believes that the [reference dose] indicated by the [National Academies] is appropriate." In its determination notice, EPA used the National Academies' reference dose—which EPA had adopted as its IRIS assessment for perchlorate in 2005—and life stage-specific exposure information to derive a range of health reference levels for 14 life stages (age groups). EPA stated in its final determination notice that it considers these potential alternative health reference levels—which range from 1 to 47 parts per billion—levels of public health concern for purposes of its final determination.

EPA Relied on an Estimate of the Relative Exposure to Perchlorate from Drinking Water and Food That It Derived from a Novel Analysis and Used a Nontraditional Method to Calculate the Relative Source Contribution

EPA conducted a novel analysis to develop estimates of exposure to perchlorate for various subpopulations that it would use to calculate the relative source contribution—the allocated exposure to perchlorate from drinking water alone. Independent scientists who reviewed EPA's analysis noted that it had several limitations—in particular, uncertainties specific to the exposure estimate for pregnant women. Nonetheless, EPA relied on the exposure estimate for pregnant women to calculate the relative source contribution, stating that the National Academies had identified pregnant women and their fetuses as the most sensitive subpopulation.[131] Further, in calculating the relative source contribution, EPA used a nontraditional method—the subtraction method—that was less conservative than the default approach it had used for its other completed regulatory determinations, and was less conservative than other available methods. While EPA identified some of the limitations of the exposure analysis in its preliminary regulatory determination notice for perchlorate, it did not discuss the effects of the limitations on EPA's exposure analysis. Moreover, although the agency's guidance for calculating the relative source contribution cautions against using the subtraction method in the absence of adequate data representative of at-risk populations—and EPA lacked data to estimate exposure to perchlorate for certain populations— the agency did not explain that the method it used to calculate the relative source contribution for perchlorate was the subtraction method or its reasoning for selecting this method.

Developing the Perchlorate Exposure Estimates

In developing its 2008 preliminary determination for perchlorate, EPA sought information on actual perchlorate exposure from food to enable EPA to estimate a "better informed relative source contribution and health reference level that is more appropriate for fetuses of pregnant women (the most sensitive subpopulations identified by the NRC [the National Academies])." EPA decided to obtain data on perchlorate exposure after it evaluated the extent of perchlorate occurrence in public water systems using the 20 percent default relative source contribution it used for its other regulatory determinations. The default relative source contribution resulted in a health reference level of 5 parts per billion. In evaluating exposure at this level, some agency officials concluded that perchlorate warranted regulation.[132] (See app. VII for more detailed information about EPA's evaluation of exposure at this level.) According to an Office of Water official, EPA had not obtained data for other regulatory determinations because the conservative default relative source contribution had not previously resulted in occurrence analyses that suggested regulation was warranted. In commenting on the exposure analysis, another Office of Water official explained that "there were too many stakeholders with high interest in the outcome" of the perchlorate regulatory determination for EPA "not to go out and get the data [it] needed."[133] This interest arose, in part, because a federal drinking water standard for perchlorate could also serve as a possible basis for environmental cleanup goals, potentially affecting some federal agencies' and perchlorate manufacturers' cleanup requirements at Superfund and other contaminated sites.[134]

The primary data sources EPA used to develop perchlorate exposure estimates were (1) CDC's National Health and Nutrition Examination Survey (CDC's biomonitoring study),[135] which included information on perchlorate exposure for a nationally representative sample of 2,820 U.S. residents, and (2) perchlorate occurrence data from EPA's testing program for drinking water contaminants. Merging CDC's biomonitoring data with perchlorate occurrence data from EPA's testing program for drinking water contaminants—a methodology EPA described as novel—EPA developed estimates of daily exposure to perchlorate from drinking water for various age and gender groups. EPA stated in its preliminary determination notice that the analysis of the merged data provides the best available information to characterize non-drinking-water exposures to perchlorate for what it identified as the most sensitive subpopulation— pregnant women and their fetuses. EPA also noted that the primary contribution of using this new exposure methodology is that the biomonitoring data can provide more direct estimates of exposure to the perchlorate that individuals consume in their diets. EPA also considered an alternative method for estimating perchlorate exposure in food that does not use biomonitoring data. In this method, perchlorate intake from food was estimated by combining two sets of estimates—FDA's estimates of perchlorate levels in different food groups[136] and USDA's estimates of food and beverage consumption—for different age and gender groups.

Although EPA's exposure analysis had several key limitations that were identified by peer reviewers, the extent to which the agency identified them and explained their significance in its preliminary regulatory determination notice was limited. Prior to its 2008 preliminary determination for perchlorate—in which EPA relied on the results of this exposure analysis—EPA had the analysis peer reviewed by three independent scientists. EPA also received comments from CDC officials, including the official who conducted the CDC biomonitoring study and coauthored, with an EPA scientist, the subsequently published 2010 article in the *Journal of Exposure Science and Environmental Epidemiology* that presented the exposure analysis methodology EPA used to support its preliminary determination for perchlorate.[137] In general, reviewers commented that EPA's exposure analysis using the merged CDC and EPA data was a novel and defensible approach to examining perchlorate exposure from food and water, but they also recognized a number of limitations. EPA identified a few of the significant limitations of its analysis from the peer review in its 2008 preliminary regulatory determination notice that was published in the *Federal Register*, but did not explain the effect of these limitations on the analysis. Further, the *Federal Register* notice did not include some important limitations of this novel methodology. Some of these limitations were raised by reviewers before EPA issued its preliminary determination, and others were subsequently identified in the 2010 journal article.[138]

On the basis of our review of the *Federal Register* notice, reviewers' comments, and the journal article, we found that the lack of a comprehensive explanation for the methodology's limitations and uncertainties reduce the credibility of EPA's use of the methodology to support its decision. Three key limitations are highlighted:

- First, because data were not available to assess the exposure to perchlorate of infants and children younger than 6 years of age,[139] EPA's analysis of perchlorate exposure using biomonitoring data does not include some age groups that are also sensitive subpopulations. Nonetheless, EPA stated that *all* sensitive subpopulations are protected by the health reference level of 15 parts per billion that the agency used as its level of public health concern in its 2008 preliminary regulatory determination for perchlorate. EPA did not identify or discuss this limitation of the biomonitoring data (e.g., lack of data on infants and children) in its presentation of the exposure analysis in the preliminary determination notice. EPA did a separate sensitivity analysis to justify the protectiveness of the health reference level for infants, as described later. The Office of Children's Health Protection noted in its comments regarding the exposure analysis that since the "real target life stage for EPA to protect are newborn infants," its usefulness may be limited. In addition, reviewers stated that infants who are breast feeding may have a much higher exposure than the populations analyzed in the study (children 6 years of age or older and adults) because, for example, data demonstrate that perchlorate in breast milk may be 10 times more concentrated than perchlorate found in urinary estimates. Further, according to a 2007 study EPA cited in its determination notice, perchlorate was measured in infant formula at levels as high as 4 parts per billion. As a result, the perchlorate dose to which formula-fed infants are exposed could be further increased if the formula is prepared with water contaminated with perchlorate. We note that, in a different section of EPA's preliminary regulatory determination for perchlorate, the agency showed that formula-fed infants can be exposed to more than twice the dose that breast-fed infants receive. However, the implications of this information were not discussed in the context of EPA's exposure analysis for perchlorate.
- Second, the exposure estimates that resulted from EPA's analysis of the CDC biomonitoring data—particularly those for pregnant women— reflect some significant uncertainties associated with inherent limitations in the method used for extrapolating an estimate of daily exposure to perchlorate from CDC's biomonitoring data. Specifically, to estimate daily exposure to perchlorate, EPA had to make an

adjustment—called a creatinine adjustment—to the CDC biomonitoring data because only one urine sample was collected from each participant—called spot testing. This adjustment is especially uncertain when made for pregnant women because of the numerous physical changes women undergo during pregnancy. The 2010 journal article on the exposure assessment states that some caution should be used when interpreting perchlorate exposure estimates for pregnant women, noting the "profound changes in a woman's body during pregnancy." In addition, a technical addendum to the journal article identifies other problematic aspects of applying the creatinine adjustment to pregnant women that make the exposure estimates "tenuous." EPA officials said they believed EPA disclosed all significant uncertainties of the exposure assessment that the agency was aware of at the time it used the assessment in supporting its preliminary regulatory determination notice. However, other reviewers that commented on the study prior to EPA's 2008 preliminary determination expressed similar concerns about the reliability of the creatinine adjustment for pregnant women. While a single statement in EPA's preliminary regulatory determination notice regarding these uncertainties reflects EPA's recognition of these issues, EPA does not explain or assess them: "A limitation is in the use of NHANES' [CDC biomonitoring study's] spot urine testing, and creatinine corrections for a population with diverse physiological characteristics to calculate the daily perchlorate dose." This statement does not identify, for example, the "population"—pregnant women—to which it is referring and does not identify the effects of these limitations on EPA's exposure analysis. This lack of disclosure is significant because of the direct impact the exposure estimates have on the relative source contribution calculation, which, in turn, directly affects the health reference level— the benchmark EPA uses to determine whether perchlorate occurs in drinking water with a frequency and at levels of public health concern.

- Third, EPA relied on an exposure estimate that the agency itself had identified as anomalous as the basis for calculating its relative source contribution. According to the principal EPA scientist conducting the analysis, EPA expected that the perchlorate exposure for individuals in the group exposed to perchlorate from food and water would be greater than from food or water alone. FDA's findings from its 2006 Total Diet Study that perchlorate is widespread in the food supply seem to corroborate this as a logical and reasonable assumption. For most of the age and gender groups, the data met this expectation. However, at both the mean and the 90th percentile, pregnant women's exposure to perchlorate from food alone was higher than for pregnant women exposed to perchlorate from both food *and* water. Specifically, in a table in EPA's preliminary regulatory determination, the agency estimated that, at the 90th percentile, pregnant women's exposure to perchlorate from food only was 0.263 micrograms per kilogram per day, while the exposure for those who ingested perchlorate through both food and water was 0.121. While EPA stated in its draft preliminary regulatory determination sent to OMB that such anomalous results added uncertainty to this exposure analysis, this disclosure was deleted during OMB's review and thus did not appear in EPA's 2008 preliminary regulatory determination notice. Some reviewers specifically questioned the analysis's conclusions regarding pregnant women's exposure to perchlorate in drinking water and stated that issues identified in their comments would need to be addressed before this approach was used in a larger context.

Further, in response to reviewers' comments, the authors of the 2010 journal article on this methodology removed four data points that were determined to be outliers—two of which were data points that corresponded to women. As a result, in the article the authors reported pregnant women's exposure to perchlorate from food alone at 0.198 micrograms per kilograms per day—compared with EPA's 0.263 estimate.

Even with this downward adjustment in the food exposure estimate for pregnant women, the anomalous relationship between exposure from food alone and from food and water likely persisted. However, readers of the journal would not be aware of such an anomaly because for pregnant women—the population of greatest significance in this study, according to EPA—the authors included only the data for those most likely to have been exposed to perchlorate through food alone and omitted data for those who were more likely to have been exposed to perchlorate through food and water. For all other populations in the study, such as children aged 6 to 11 and women aged 15 to 44, the authors report all exposure data. Appendix VIII provides greater detail on these and other limitations of EPA's perchlorate exposure analysis. In addition, we found that while the downward adjustment in the food exposure estimate for pregnant women in 2010 resulted in a relative source contribution estimate for perchlorate exposure from water of 72 percent compared with the 62 percent relative source contribution derived for EPA's preliminary determination in 2008, the 2008 and 2010 analyses both specifically support the same health reference level of 15 parts per billion—the level EPA and the Perchlorate Interagency Working Group agreed to. Appendix IX shows how making small changes in the standard assumptions for weight and drinking water intake that were used to calculate the health reference levels in 2008 and 2010 allowed EPA and the journal authors to maintain consistency in the health reference levels reported.

Calculating the Relative Source Contribution

Notwithstanding the limitations of the exposure analysis discussed earlier, EPA used the exposure estimate for pregnant women to calculate a relative source contribution for perchlorate using a nontraditional method—that is, the subtraction method—rather than applying the default 20 percent relative source contribution it has used for its other regulatory determinations. EPA officials said they used the subtraction method for perchlorate because they had the data to estimate exposure for perchlorate, whereas they had not for other contaminants. As previously discussed, however, according to EPA's guidance on methods for developing relative source contribution estimates,[140] if the agency determines that it has adequate exposure data to estimate the relative source contribution, rather than apply the default assumption, either of two methods may be applicable—the percentage method or the subtraction method.

EPA's decision to use the subtraction method rather than the percentage method to calculate the relative source contribution had a significant effect on the resulting health reference level, and consequently on EPA's evaluation of whether (1) perchlorate occurred at a frequency and at levels of public health concern and (2) the regulation of perchlorate would provide a meaningful opportunity for public health risk reduction. Significant differences in the resulting health reference levels can occur because the subtraction method allows health reference levels to be set at the highest levels short of exceeding the reference dose. In contrast, the percentage method allows for a cushion between actual exposure levels and the reference dose, allowing for other sources of exposure and risk assessment uncertainties, including uncertainties related to the exposure data. For example, using EPA's exposure estimates for people 20 years of age or older,[141] the health reference level for perchlorate would be 19.6 parts per billion using the subtraction method.[142] In contrast, the health reference level would be 5 parts per billion using the percentage method.[143] The sensitivity of the resulting health reference level—which can drive policy decisions—underscores the need for transparency in EPA's regulatory determination notices regarding the relative source contribution methodology the agency selects.

EPA's guidance on relative source contribution methods indicates that the subtraction method may be used in some cases in which data are available—relevant to and representative of populations at risk—to describe central tendencies and high-end exposure. Further, the guidance generally advises that the subtraction method should be used with caution because it removes any cushion between existing exposure levels and the reference dose.[144] An Office of Water official said that it is extremely rare for EPA to have the exposure data it needs to use the subtraction method.

Because EPA did not document a determination that it had sufficient data, it is not clear why EPA deemed it appropriate to use the subtraction method to calculate the relative source contribution for perchlorate in light of the exposure analysis's lack of data on infants—another sensitive subpopulation identified by the National Academies and EPA—as well as the uncertainties in EPA's exposure analysis of pregnant women. In November 2010, Office of Water officials told us that they believe the use of available exposure information using the subtraction method for relative source contribution was consistent with the requirement in the Safe Drinking Water Act to use "the best available, peer-reviewed science and supporting studies conducted in accordance with sound and objective practice." Office of Water officials also said that EPA's 2000 guidance on when to use the percentage and subtraction methods "does not reflect more current thinking."[145] We note, however, that EPA's 2008 preliminary regulatory determination notice on perchlorate did not expressly disclose (1) that the agency was using the subtraction method, (2) that this was the first use of this method in a preliminary (or final) regulatory determination, or (3) any guidance or rationale for the use of the subtraction method. Office of Water officials told us that the choice of relative source contribution methodology was either using the 20 percent default assumption when exposure information is not available or using the subtraction method in cases where information is available to develop estimates of exposure to drinking water contaminants from multiple sources. However, without EPA documentation, it remains unclear whether this position is consistent with the Office of Water's existing guidance on relative source contribution, which provides that there may be a choice between the use of the percentage or subtraction method— depending, in part, on the scope and quality of the exposure data.

Throughout the time that the agency was considering its course of action regarding a relative source contribution for perchlorate, the small group of high-level EPA officials—in accordance with the Perchlorate Interagency Working Group's "no-surprises policy"—was sharing preliminary information and analyses with OMB and other members of the Perchlorate Interagency Working Group, and receiving direction and feedback from OMB. For example, documents we reviewed indicated that OMB officials, along with EPA management, directed the Office of Water to obtain additional data on sources of perchlorate exposure in order to estimate the relative source contribution rather than rely on the more conservative 20 percent default. In addition, the documents show that, as early as 2006, OMB had concluded that the exposure data EPA was in the process of obtaining and analyzing would support a determination not to regulate perchlorate. Further, according to these documents, OMB expected that EPA would be able to strengthen its case for not regulating perchlorate by 2007, when the agency was expected to make its second cycle of regulatory determinations.

According to Key EPA Scientists, the Agency Mischaracterized Important Scientific Findings That Emerged from Its Novel Analysis of the Sensitivity of Various Age Groups to Perchlorate in Drinking Water

In early 2008, EPA used a physiologically based pharmacokinetic (PBPK) model[146] to (1) evaluate the relative sensitivity of sensitive subpopulations to the health reference level the agency had developed based on pregnant women and their fetuses and (2) address concerns that some sensitive subpopulations, such as infants, exposed at the health reference level may receive concentrations of perchlorate above the reference dose.[147] For its preliminary regulatory determination, the agency used the model in a novel way and, according to some key EPA scientists, mischaracterized the findings of the modeling analyses by selecting and presenting information in such a way as to support the agreed-upon conclusion that a health reference level of 15 parts per billion was protective of *all* sensitive subpopulations, including infants.[148]

Specifically, in January 2008, EPA's Office of Water asked the Office of Research and Development to review the PBPK model—which had recently been used by other scientists to estimate the effects of exposure to perchlorate and how the body processes it149, [150]—to determine whether (1) that model represented the best available science and (2) there were scientifically defensible applications of the PBPK model for EPA's purpose of assessing the risks of exposure to perchlorate. Office of Research and Development officials told us that this was different from how PBPK models are traditionally used because in this case, EPA was using the model to estimate an effect—that of perchlorate exposure on inhibition of iodide uptake on the thyroid at specific drinking water concentrations— rather than to estimate the amount of perchlorate (the dose) that would adversely affect the thyroid. After examining the model in detail and making some minor adjustments to it, the Office of Research and Development concluded that it was appropriate to use the PBPK model for assessing exposure risks. Over the next several months, the Office of Research and Development used the model to conduct numerous sensitivity analyses for sensitive subpopulations, applying various sets of assumptions. As it conducted this work, the Office of Research and Development kept the Office of Water apprised of its results, some of which indicated that to prevent some infants from receiving a perchlorate dose greater than the reference dose, the concentration of perchlorate in water should not exceed 2 parts per billion. During this time, members of EPA's small group of high-level officials continued to meet with the Perchlorate Interagency Working Group, and, in early July 2008, the Assistant Administrator of the Office of Research and Development told his staff that the Perchlorate Interagency Working Group had requested a particular analysis using the PBPK model. Specifically, the Perchlorate Interagency Working Group asked EPA to use the model to show the iodide uptake inhibition rate for each sensitive subpopulation at water concentrations of 20 and 15 parts per billion. According to the Assistant Administrator, this was an effort to explore ways the PBPK model could be used to evaluate the effects of perchlorate exposure "instead of using default approaches."

EPA presented the results of such a PBPK analysis in its October 2008 preliminary regulatory determination for perchlorate to support its conclusion that a health reference level of 15 parts per billion was protective of all sensitive subpopulations, including infants, and stated that using the model in this way could reduce some of the uncertainty regarding the sensitivities of subpopulations other than pregnant women. The analysis EPA presented in its preliminary regulatory determination compared the (1) estimated iodide uptake inhibition rate predicted by the model for various age groups with the (2) iodide uptake inhibition rate corresponding to the no-observed-effect level for the seven healthy adults from which the reference dose was derived (point of departure), at the health reference level of 15 parts per billion. From this comparison, the agency concluded that the 2.2 percent iodide uptake inhibition rate estimated by the model for the formula-fed infant—the age group with the highest inhibition rate—was comparable to the 1.8 percent inhibition rate for a healthy adult, noting that the difference in the inhibition rates was "within the statistical uncertainty in the data."

According to Office of Research and Development officials overseeing this work, however, this comparison was "misleading and incomplete" because it does not account for the uncertainty and variability inherent in comparisons of the responses of adults with those of other age groups, such as infants and children. These officials also stated that the comparison in the preliminary determination notice of the model-predicted iodide uptake inhibition to the iodide uptake inhibition associated with the point of departure[151] identified by the National Academies was inappropriate. They also said that a comparison of the model-predicted iodide uptake inhibition would also need to be made with the iodide uptake inhibition associated with the reference dose to understand whether a drinking water concentration of 15 parts per billion would be protective of all subpopulations. Further, Office of Research and Development officials said they disagreed with the way EPA presented the information in its preliminary regulatory determination notice, saying the agency did not sufficiently explain the uncertainties and limitations, instead presenting the information more conclusively than was appropriate. In discussing these issues regarding the comparison with Office of Water officials, however, they said the comparison was valid on the basis that EPA agrees with the National Academies identification of iodide uptake inhibition as a valid measure to protect against the potential for an adverse health effect from perchlorate.[152]

While EPA relied on the modeling results and comparison discussed earlier to support its conclusion that a health reference level of 15 parts per billion was protective of all subpopulations, the table EPA published in the preliminary regulatory determination notice presenting the results of this analysis also included data that may not be consistent with that conclusion. That is, while the table providing the estimated daily dose of perchlorate to which infants and children up to age 2 would be exposed at a drinking water concentration of 15 parts per billion did not present a comparison of these estimated exposures with the reference dose, the table did provide sufficient data for informed readers of the preliminary determination to calculate such comparisons. Such calculations show that infants and young children could be exposed to doses of perchlorate at levels as high as 5.5 times greater than the reference dose, supporting the concern that infants and young children may, in fact, be more vulnerable to perchlorate exposure. Moreover, while EPA stated in the notice that "for some [subpopulations], the modeled exposure exceeds the [reference dose]," the agency was not explicit about the extent to which the reference dose is exceeded—as calculated above—and did not explain the implications of this result on its conclusion that the health reference level of 15 parts per billion is protective of all subpopulations. In providing comments on the draft notice to the Office of Water, an Office of Research and Development scientist noted that the agency's failure to present a comparison of the estimated daily exposure with the reference dose constituted a "serious omission," and characterized the infants' estimated exposure as "substantially higher" than the reference dose.

As previously mentioned, in addition to the modeling analysis discussed above, the Office of Research and Development had conducted numerous other sensitivity analyses with the PBPK model that were not presented or discussed in EPA's October 2008 preliminary regulatory determination for perchlorate. Some of the analyses were subsequently presented in a May 2009 EPA report on the sensitivity analyses following an independent peer review and public comment period of the analyses.[153] One of the analyses included in the report was an explicit comparison between estimated perchlorate doses received by all population groups and the existing reference dose that a senior EPA scientist had identified in September 2008 (a month before the preliminary determination was published) as "necessary for an adequate health risk characterization for perchlorate exposures from drinking water." According to an Office of Research and Development official, another analysis included in the report that compared the model-predicted iodide uptake inhibition of sensitive subpopulations with the model-predicted iodide uptake inhibition of an average adult at the no-observed-effect level dose identified by the National Academies was more appropriate than the analysis presented in the *Federal Register* notice. In this comparison, the PBPK model estimated an iodide uptake inhibition rate for the age group with the highest inhibition rate—again the 7-day-old formula-fed infant—of 4.3 percent, compared with a rate of 1.6 percent for an average adult. In contrast, the PBPK analysis EPA used for its preliminary regulatory determination predicted a 2.2 percent iodide uptake inhibition for a 7-dayold formula-fed infant that EPA concluded was "comparable" to the 1.8 percent iodide uptake inhibition that the National Academies recommended as the point of departure for calculating the reference dose. This analysis showing a greater disparity between the iodide inhibition rates in adults and infants was available before EPA issued its preliminary determination.

EPA's limited presentation of the numerous PBPK analyses conducted by the Office of Research and Development in its preliminary regulatory determination notice validated the concern expressed at the time by Office of Research and Development scientists who conducted the analyses: that individual analyses could be used out of context in a way that could be misleading. Accordingly, Office of Research and Development scientists conducting the PBPK analyses said it was important to present the different analyses in the interest of informed and transparent decision making. An Office of Research and Development official articulated in September 2008 that while his office and the Office of Water had developed careful and sophisticated PBPK analyses to support the agency's preliminary regulatory determination, "the use of these science results in [the] draft regulatory determination is seriously flawed and misleading." The official further noted that the draft interpreted these analyses indicating the absence of a perchlorate health risk for infants, but that the results actually indicated that infants would receive a perchlorate dose "approximately an order of magnitude greater than the reference dose." As a result, Office of Research and Development officials and scientists that conducted the analyses concluded that the PBPK analysis done by the office did not support the draft preliminary regulatory determination's suggested health reference level of 15 parts per billion as being health protective for all sensitive subpopulations of concern to EPA. While these Office of Research and Development officials and scientists communicated their disagreement regarding the fundamental implications of, and the agency's presentation of the PBPK modeling results in the draft preliminary regulatory determination to the Office of Water—which was managing revisions to the notice as it underwent OMB review—they were not given an opportunity to review the final draft of the preliminary determination notice, and some key changes that they suggested were not reflected in the notice that was published in October 2008.

EPA stated in its preliminary determination notice that it planned to issue a final determination by December 2008. In August 2009, however, instead of finalizing its 2008 preliminary regulatory determination after evaluating public comments—which is the typical practice—EPA decided to seek a second round of public comments through a *Federal Register* notice, specifically requesting input on sensitive populations. In this notice, EPA indicated that it was reevaluating how to best incorporate PBPK modeling into its evaluation of perchlorate, if at all, and presented an alternative PBPK analysis discussed earlier that explored the relative sensitivity of the various life stages to a fixed dose (the no-observed-effect level dose)—an analysis that EPA scientists who conducted the PBPK modeling believed should have been included in the 2008 preliminary regulatory determination in order to present the science correctly. This analysis showed, for example, that the model-predicted iodide uptake inhibition is approximately one and one-half-fold higher for the formula-fed infant compared with the average adult. In addition, EPA reported in this notice that infants less than 6 months in age generally consume five to eight times more water than pregnant women or women of child bearing age on a per body weight basis, and thus will receive a higher dose for any given drinking water concentration. EPA stated in its February 2011 final determination to regulate perchlorate that it is evaluating the alternative health reference levels it had developed in its August 2009 notice—based on exposure information for 14 life stages—and considers them to be levels of public health concern for purposes of its regulatory determination.

EPA's Independence in Developing and Communicating Its Scientific Findings Was Limited by Its Acceptance of External Input on the Preliminary Determination Notice

Compounding scientists' concerns about the mischaracterization and lack of transparency regarding relevant scientific analyses, key language in EPA's preliminary regulatory determination notice appears to have been drafted by OMB rather than EPA. In working to finalize the preliminary regulatory determination notice, EPA's Office of Water worked with OMB, whose clearance of the notice was required per EPA's policy implementing Executive Order 12866 before the Office of Water could provide it to the EPA Administrator for review, approval, and publication in the *Federal Register*.[154] According to the Office of Water, in four iterations of review, OMB sent EPA a substantial number of comments on the notice; in response, EPA "clarified its description of the supporting analysis and strengthened the rationale for the determination." The following example highlights OMB's role in reviewing and approving the specific wording of EPA's scientific analyses regarding perchlorate exposure in infants and children:

> Text EPA provided to OMB: "Because infants and children eat and drink more on a per body weight basis than adults, eating a normal diet and drinking water with 15 [micrograms per liter] of perchlorate *is likely to* result in exposure that is greater than the reference dose in these groups."
>
> Revised text provided to EPA by OMB: "Because infants and children eat and drink more on a per body weight basis than adults, eating a normal diet and drinking water with 15 [micrograms per liter] of perchlorate *may result in* exposure that is greater than the reference dose in these groups."

By changing three words, OMB downplayed EPA's characterization of the health risks of perchlorate exposure. Importantly, the EPA scientist who wrote the text provided to OMB noted to EPA reviewers—before it was sent to OMB in August 2008—that the PBPK model actually showed exposures at levels "much higher" than the reference dose, but also said that he believed describing the exposure scenario as "likely" was the strongest characterization that might be retained through OMB review. In addition, in September 2008, during its review of the draft preliminary determination notice and before clearing it for publication, OMB reminded EPA that it expected the notice to "state a clear conclusion that the HRL [health reference level] is protective of all subpopulations, as agreed to in the August framework"—and accordingly, this conclusion appeared in the agency's October 2008 preliminary determination notice.[155]

Overall, the changes EPA made in response to OMB's extensive comments through the external review process downplayed the health risks of exposure to perchlorate and presented EPA's conclusions with greater certainty than key EPA scientists stated they were comfortable supporting. In addition, it is difficult to identify the nature and extent of OMB's changes to the 2008 preliminary regulatory determination notice on perchlorate because OMB provided four iterations of comments, resulting in substantial revisions; EPA did not make OMB's actual comments on the regulatory determinations part of the public record. Rather, an Office of Water official developed a statement for the record briefly summarizing the nature and extent of the comments. EPA included a copy of this statement, the draft determination notice that was sent to OMB for clearance, and a copy of the final version of the draft determination notice for publication in the public record. We note that the approach used by the Office of Water to make OMB's comments on regulatory determinations part of the public record is different from that used in other programs, such as the IRIS program, which makes actual comments on draft assessments provided by OMB and other federal agencies available to the public. This latter approach is more transparent and supports a clear understanding of the nature and extent of OMB's proposed changes. Moreover, the practice of fully disclosing individual comments promotes transparency, which is one of EPA's core values established in its strategic plan, and which the Administrator emphasized in a 2009 memorandum on the importance of transparency sent to all EPA employees.[156] The memorandum established transparency guidelines intended to maintain the fairness and openness of EPA operations and thus strengthen public confidence in EPA decisions.[157]

CONCLUSION

Enacted with the goal of assuring the safety of public drinking water, the Safe Drinking Water Act authorizes EPA to regulate contaminants in public drinking water systems. Since the enactment of the 1996 amendments to the act, EPA had not recommended any new contaminants for regulation until February 2011, when it, among other things, reversed its 2008 preliminary decision to not regulate perchlorate. Systemic limitations in its implementation of the 1996 amendments' requirements for determining whether additional contaminants in public drinking water warrant regulation have impeded EPA's progress in helping assure the public of safe drinking water. Specifically, EPA lacks (1) criteria and a process for identifying those contaminants on its candidate list that pose the greatest public health concern and (2) a coordinated process to obtain occurrence and health effects information to support informed regulatory determinations, resulting in regulatory determinations on contaminants selected on the basis of available data rather than on the basis of public health concern.

In addition, while the unregulated contaminants testing program mandated by the 1996 amendments provided EPA with the authority and a structure to obtain consistent, high-quality occurrence data that could effectively support regulatory determinations on individual contaminants, the extent and utility of the occurrence data this program generates to support EPA's drinking water program is reduced because the agency has not (1) taken full advantage of its statutory authority to require testing for the maximum number of contaminants allowed, 30, under each 5-year testing cycle, and (2) used sufficiently sensitive analytic tests in some cases. We recognize that EPA may not always know the health reference level for a contaminant at the time the analytic method is being selected because a health risk assessment may be in progress. However, until the agency resolves this timing challenge—such as by the use of the more sensitive tests USGS has used for many years—EPA may be unable to routinely develop occurrence data at appropriately low detection levels to support informed regulatory determinations. Also, EPA's decision to use the more limited screening survey for a majority of the contaminants tested, which can provide national occurrence data with greater uncertainty than those provided by assessment monitoring, has to date limited the utility of some of the testing data obtained, and it remains unclear whether EPA will find these data sufficient for future determinations.

Importantly, EPA's Office of Water has not developed policies and guidance that can help agency staff interpret and apply the 1996 amendments' broad statutory criteria, or developed a protocol to follow in making regulatory determinations that can both guide staff and incorporate accountability into the process. Without such policies and guidance, the basis for EPA's regulatory determinations—and the quality of the documentation the agency uses to support them—can fluctuate over time with changes in agency leadership and staff. Also, without guidance, EPA lacks thresholds or parameters for applying the broad statutory criteria to ensure their consistent application. For example, EPA is informally applying an unspecified national occurrence requirement for contaminants to meet the statutory criterion of occurring "with a frequency and at levels of public health concern." This national occurrence requirement has not been defined or vetted internally or externally, yet its use represents a critical interpretation of the Safe Drinking Water Act that has important implications for the safety of public drinking water in the United States. Moreover, without guidance on such things as the characteristics of occurrence data and health effects information that are adequate to support regulatory determinations, methods and analyses to evaluate the health effects on sensitive subpopulations, and a process to ensure that the presentation of occurrence data and health effects information in regulatory determination notices and support documents is comprehensive, consistent, informative, and understandable, EPA's support for its regulatory determinations will continue to lack transparency, clarity, and consistency—affecting their credibility. For example, such information cannot be comprehensive without clear explanations of key information, such as whether and how EPA used various data, the relative source contribution method the agency used to calculate the health reference level, instances in which the minimum reporting levels for data used in assessing contaminants' occurrence in drinking water were not sufficiently sensitive to detect occurrence at the level of public health concern, and any exceptions to existing guidance reflected in the agency's support for its regulatory determinations. In addition, as illustrated by EPA's preliminary determination on perchlorate, information on the relative source contribution method the agency uses to calculate the health reference level is also key to providing a comprehensive understanding of regulatory determinations. Moreover, in light of the importance of the regulatory determinations to ensure the safety of public drinking water and protect public health, the lack of guidance hampers EPA's ability to consistently apply and document (1) the statutory requirement to consider sensitive subpopulations—such as infants and children, those with kidney and liver disease, those with compromised immune systems, and the elderly—when selecting contaminants for regulatory determination, and (2) EPA's children's health policy, which specifies how health risks to infants and children should be analyzed when the agency is developing a policy, regulatory action, or decision. As demonstrated by the preliminary determination for perchlorate, EPA also had not determined how to use the expertise of officials and scientists in relevant offices, including the Office of Children's Health Protection, in making decisions regarding drinking water contaminants, such as by ensuring that, to the extent possible, such offices consistently participate in regulatory determination work groups.

In the absence of guidance defining the circumstances and process under which EPA may want to reconsider final determinations to not regulate, it is unclear whether and to what extent EPA would consistently do so. In addition, in issuing policies and guidance, EPA could be better assured that the guidance would address identified shortcomings by having an independent review of any proposed guidance by an entity with scientific and technical credibility, such as either of EPA's two standing drinking water advisory committees. Importantly, guidance alone may not ensure that EPA conducts future regulatory determinations in a consistent manner without an internal review mechanism to ensure the application of any guidance. Such a mechanism could be instrumental in preventing the use of an atypical process such as the one used for the preliminary determination on perchlorate—which lacked transparency and limited the agency's independence in developing and communicating its scientific findings.

EPA has posted health advisories on its Web site to inform states, localities, and public water systems that action on their parts may be required to protect public health from exposure in public drinking water to 9 of the 20 contaminants the agency has decided not to regulate. However, because EPA does not require public water systems to monitor for these unregulated contaminants, current data on the occurrence of these contaminants in the nation's public water systems are generally not available. The lack of actual comprehensive data on the occurrence of these contaminants in public water systems, widespread state and local government budget constraints, and, in some cases, limitations in states' ability to require systems to conduct testing, are factors that could hamper efforts by states and localities to use the health advisories in a timely and effective manner to protect public health. Further, while not directed to consumers, the health advisories in some cases contain warnings about using public water to prepare formula and food for infants and children containing the contaminants at certain levels. Because many consumers would have to find and pay a laboratory to have their water tested to be informed of the occurrence of these contaminants in their drinking water, it would also be difficult for them to effectively use the advisories to protect their health and that of their families.

Finally, EPA currently submits its regulatory determinations to OMB for review and clearance pursuant to executive order and OMB policy, but for the 2008 preliminary regulatory determination on perchlorate—and for the prior 20 determinations—OMB's individual comments and direction are not transparently identified in the public record. This information is key to understanding the degree to which EPA is maintaining its independence in developing and communicating the results of its scientific analyses. Moreover, the lack of transparency of OMB's overall review and clearance is important because, for example, in discussing complex scientific analyses and conclusions, changes that appear to be minor can actually significantly alter the conclusions presented in the determinations— conclusions can be made to appear more certain than they actually are or, conversely, can be made to appear less important than EPA scientists assess them to be.

RECOMMENDATIONS FOR EXECUTIVE ACTION

To increase EPA's consistency, transparency, and clarity in implementing the Safe Drinking Water Act in a way that better assures the public of safe drinking water, we recommend that the Administrator of EPA take the following 17 actions.

To systematically implement the statutory requirement to consider for regulation the contaminants that present the greatest public health concern, we recommend that the EPA Administrator require that the Office of Water

- develop criteria and a process for identifying those contaminants on its candidate list that present the greatest public health concern, and
- develop a coordinated process for obtaining both the occurrence and health effects data that may be needed for the agency to make informed regulatory determinations on these priority contaminants.

To take full advantage of the opportunities provided by the testing program mandated by the statute and thereby obtain high-quality occurrence data on the authorized number of unregulated contaminants, we recommend that the EPA Administrator require the Office of Water to take the following steps:

- Use its full statutory authority to test for the 30 contaminants allowed under each 5-year testing cycle.

- Conduct testing for most or all of the selected contaminants using the assessment monitoring program, rather than the more limited screening surveys, to obtain robust occurrence data, from which national estimates with high confidence levels can be derived.
- Select minimum reporting levels for testing selected unregulated contaminants that are sufficiently sensitive to reliably (1) detect the known and likely occurrence of contaminants in public water systems at levels of public health concern and (2) provide useful and credible information on the occurrence of the contaminants in public drinking water systems.

To support the development of regulatory determinations that are transparent, clear, and consistent and that follow applicable agency policy, we recommend that the EPA Administrator require the Office of Water to expeditiously develop, and make available to the public, policies or guidance that clearly articulates the agency's interpretation of the act's broad statutory criteria for making regulatory determinations and provides a protocol for making such determinations.

In particular, the guidance should

- specify any thresholds or parameters that the agency requires to be met to support a positive finding for each criterion to ensure their consistent application.
- include factors for determining when the occurrence and health effects data the agency identifies are adequate to support a regulatory determination.
- establish a process to ensure that the presentation of health effects and occurrence information in regulatory determination notices and support documents is comprehensive, consistent, informative, and understandable and that it includes clear explanations of key information, such as
  - whether and how EPA used various data,
  - the relative source contribution method the agency used to calculate the health reference level,
  - instances in which the minimum reporting levels for data used in assessing contaminants' occurrence in drinking water are above the health reference level (e.g., are not sufficiently sensitive to detect occurrence at the level of public health concern) and the limitations of using such occurrence data to support regulatory determinations, and
  - any exceptions to existing guidance reflected in the agency's support for its regulatory determinations.
- establish the approaches, such as methods and analyses, as appropriate, to evaluate the health effects on sensitive subpopulations, including such groups as infants and children, those with kidney and liver disease, those with compromised immune systems, and the elderly, and to comply with applicable agency policy and guidance for assessing children's health risks.
- specify that appropriate stakeholders—that is, EPA offices with relevant expertise such as the Office of Children's Health Protection and regional offices that have known or likely occurrence of the contaminants being evaluated in public water systems within their areas of jurisdiction—be encouraged and have the opportunity to participate in the regulatory determination work groups.

- define the circumstances under which, and the process EPA will use, to reconsider whether to regulate a contaminant for which it previously issued a determination not to regulate and, in the context of the recommended guidance, consider whether the agency needs to reevaluate any of its past determinations to not regulate.

We further recommend that a draft of the guidance we are recommending EPA develop be reviewed by the Science Advisory Board's Drinking Water Committee or the National Drinking Water Advisory Council and that EPA consider the committee's comments before finalizing the guidance. In addition, we recommend that the EPA Administrator develop and implement an internal review mechanism to help ensure that EPA's regulatory determinations are consistent with the guidance.

In light of EPA's decisions to issue health advisories in conjunction with determinations to not regulate certain contaminants that have been detected in some public water systems at levels of public health concern, we recommend that the EPA Administrator (1) determine whether the Office of Water's use of health advisories provides sufficient information on these unregulated contaminants to support timely and effective actions by states, localities, public water systems, and the public to ensure the safety of public drinking water, and (2) if not, direct the Office of Water to develop a plan to more effectively communicate such information to these entities.

To improve transparency and help EPA ensure that it maintains the fairness and openness of its operations and thus strengthens public confidence in its decisions, we recommend that the EPA Administrator require the Office of Water to include in the public record OMB's and other federal agencies' comments on and revisions to regulatory determination notices and support documents.

AGENCY COMMENTS AND OUR EVALUATION

We provided a draft of this report to the Administrator of EPA for review and comment. In commenting on the draft report, EPA did not say whether it agreed or disagreed with our findings. EPA agreed with two of our recommendations regarding its drinking water health advisories, stating that it would evaluate their utility and determine whether and how to revise the advisories to better serve states, localities, public water systems, and the public. However, EPA did not agree to adopt the remaining 15 recommendations we made to improve EPA's implementation of the Safe Drinking Water Act's requirements for determining whether additional drinking water contaminants warrant regulation. These recommendations are largely aimed at developing basic guidance to support clear and consistent agency actions to help assure the public of safe drinking water.

Regarding our recommendation to develop criteria and a process for identifying those drinking water contaminants on its candidate lists that present the greatest public health concern, EPA took the position that no action is needed, as the agency's candidate list itself represents the contaminants of greatest public health concern. We acknowledge in our report that the candidate list represents one level of prioritization, but we also identify the need for EPA to further prioritize the 116 contaminants on its current candidate list to obtain coordinated and timely health effects and occurrence data and make regulatory determinations on contaminants of greatest public health concern. We also report that in 2009, EPA's Science Advisory Board concluded that the candidate list is too large to effectively prioritize contaminants for the purpose of making regulatory determinations. Especially in light of increased fiscal constraints and programmatic demands, we believe it is of utmost importance for EPA to prioritize the contaminants on its candidate list to identify those of greatest public health concern so that the agency can effectively target its limited resources to obtain the data it needs to support regulatory determinations on priority contaminants. EPA noted that the agency "is prioritizing" contaminants from the candidate list for the next cycle of regulatory determinations. The implication of this statement is that EPA will, in fact, prioritize contaminants, but without having established criteria and a process for so doing. We continue to believe that establishing criteria and a process for identifying contaminants on the candidate lists that present the greatest public health concern is important to provide transparency and credibility to EPA's implementation of the statutory requirement for prioritizing.

Regarding our recommendation to develop a coordinated process for obtaining the occurrence and health effects data needed to make informed regulatory determinations on the contaminants of greatest public health concern, EPA's response describes its current efforts for obtaining these data—such as engaging in "internal and external discussions about data gaps that need to be filled" and participating in the Office of Research and Development's planning process to prioritize health effects research for the next 5 years—and does not acknowledge that any further steps are needed. As we reported, the approaches EPA has used, and advocates continued reliance on, have resulted in the agency's making limited progress in filling data gaps since the 1996 amendments. Moreover, the data gaps for contaminants on its current candidate list are substantial— EPA identified a lack of sufficient occurrence or health effects data, or both, for at least 100 of the 116 contaminants on the list. Therefore, we continue to believe that the Office of Water needs to develop a coordinated process for obtaining both occurrence and health effects data for those contaminants on its candidate list that present the greatest public health concern.

EPA did not agree to implement our recommendations that the EPA Administrator require the Office of Water to fully and effectively utilize the testing program by (1) using EPA's full statutory authority to test for the 30 contaminants allowed under each 5-year testing cycle; (2) conducting testing for most or all of the selected contaminants using the assessment monitoring program; and (3) selecting minimum reporting levels that are sufficiently sensitive to, for example, reliably detect the occurrence of contaminants in public water systems at levels of public health concern. While we understand from its response that EPA currently supports fully utilizing this important program, EPA has not done so in the past; we believe it is important for EPA to implement our recommendations and thereby institutionalize the agency's commitment to fully and effectively using the unregulated contaminants testing program going forward.

To support the development of regulatory determinations that are transparent, clear, and consistent and that follow applicable agency policy, we made an overarching recommendation that EPA develop policies or guidance that clearly articulate the agency's interpretation of the Safe Drinking Water Act's broad statutory criteria, as well as eight additional recommendations identifying specific components of this guidance and calling for review of the draft guidance by one of EPA's independent advisory committees and the establishment of an internal review mechanism to help ensure the determinations are consistent with the guidance. EPA did not agree to implement these nine recommendations. Specifically, EPA said it believed that establishing policies or guidance for regulatory determinations was not "practicable" because of the many combinations of health effects factors and potential ranges of frequencies and levels of contaminants measured in drinking water. We do not believe that the existence of variables or complexities is a basis for not developing guidance for EPA staff to implement the statutory requirements for regulatory determinations. In fact, the complexities cited would argue for, rather than against, the need to develop guidance for staff on applying the criteria. EPA also did not agree with these recommendations on the basis that policies or guidance could "inhibit its ability to continually improve its actions." This perspective suggests that guidance *per se* lacks flexibility. We do not agree that guidance and flexibility are incompatible or that developing guidance would inhibit EPA's ability to improve its actions. Rather, flexibility can and should be incorporated into guidance by establishing parameters or options for areas in which flexibility is deemed appropriate. Moreover, consistency and accountability are lacking in this important program because EPA has not developed guidance on the application of the broad statutory criteria, which are susceptible to varying interpretations. As stated in our report, the statutory criteria are so broadly stated that they could potentially be interpreted so as to lead to regulating all the contaminants on the candidate list, some of them, or none of them. In its comments, EPA highlighted that, under these criteria, ultimately it is the Administrator's judgment as to whether regulation of a contaminant in drinking water presents a meaningful opportunity for health risk reduction, after considering the information presented by agency staff. It is precisely for this reason that we believe it is essential for the staff to have sufficient guidance on applying the broad criteria consistently and transparently so that the Administrator's judgment can be based on sound and consistent information. Without such guidance, the basis for EPA's determinations and the quality of the documentation the staff use to support them can fluctuate over time as a result of, among other reasons, changes in agency leadership and staff.

Regarding our recommendation that EPA develop guidance that would specify any thresholds or parameters that the agency requires to be met, EPA stated that there is not a "one size fits all" model for evaluating contaminants for drinking water regulation. We agree, but we do not believe that this precludes the development of useful and important guidance, particularly with regard to interpreting contaminant occurrence "with a frequency and at levels of public health concern" that would provide "a meaningful opportunity for health risk reduction." We believe EPA should be able to conceptualize the key scenarios that could reasonably meet these criteria, as well as anticipate, for various situations, the types of analyses appropriate to the evaluation. In our report, we cited a 1993 EPA report that provides some potential scenarios for regulating drinking water contaminants. As we said, this report could serve as a starting point in developing this important guidance. Moreover, if in fact EPA does not believe any thresholds or parameters are applicable to its regulatory determinations, the agency should expressly state this in guidance. The Safe Drinking Water Act does not require that contaminants be found in public water systems on a national basis for an Administrator to find a meaningful opportunity for health risk reduction; it also provides for relief from monitoring and for flexibilities for instances in which a contaminant occurs in certain areas but not in others. As we reported, EPA's informal use of a national occurrence requirement has significant implications for the safety of public drinking water, but EPA has not defined or vetted this critical interpretation of the statutory criterion internally or externally.

Regarding our recommendations for EPA to (1) develop guidance that includes factors for determining when the occurrence and health effects data the agency identifies are adequate to support a regulatory determination and (2) establish a process to ensure that the presentation of health effects and occurrence information is comprehensive, consistent, informative, and understandable, and that it includes clear explanations of key information we outlined, EPA stated that it will strive to improve the clarity and transparency of its support documents and *Federal Register* notices so that the public better understands the information and data being considered, any potential limitations or exceptions, how it uses this information to make decisions, and the rationale for its decisions. However, we continue to believe that the guidance we recommend is integral to EPA's accomplishing and sustaining the improvements it recognizes are needed.

Regarding our recommendation to establish approaches, such as methods and analyses, to evaluate the health effects on sensitive populations— including such groups as infants and children, those with kidney and liver disease, those with compromised immune systems, and the elderly—and to comply with applicable agency policy and guidance for assessing children's health risks, EPA states it will consider how to incorporate the elderly into its evaluations and to continue to coordinate with the Office of Children's Health Protection to improve the methods it uses to assess health effects for regulatory determinations. These actions are not responsive to our recommendation and provide no assurance that EPA will not continue to assess sensitive populations in a manner that lacks transparency and internal consistency.

In response to our recommendation that additional guidance is needed to specify that appropriate stakeholders be encouraged to, and have the opportunity to, participate in the regulatory determination work groups, EPA stated it did not believe additional guidance is needed, opting to continue to rely on its existing administrative process (the Action Development Process) to fill this role. However, we believe that in some cases a more proactive approach is needed to ensure that appropriate stakeholders—EPA offices with relevant expertise and regional offices in areas with known or likely occurrence of the contaminants being evaluated—participate in the regulatory determination work groups. As we reported, officials from the Office of Children's Health Protection said they were involved only informally and intermittently in the first two regulatory determination work groups largely as a result of staff and resource limitations. As a result, in some cases, the Office of Water may need to augment the voluntary approach of the administrative process by identifying relevant stakeholders and working with their management, if need be, to encourage their participation.

Furthermore, EPA stated that it does not believe guidance is needed to define the circumstances under which it will reconsider whether to regulate a contaminant it has determined not to regulate, and the process EPA will use to do so, because its candidate list process allows for reevaluation of contaminants for which EPA has previously made negative regulatory determinations if new health effects or occurrence information becomes available. We believe good management necessitates specific guidance on how and when the agency will reconsider determinations to ensure that this is done consistently and in a timely manner. Also, we continue to believe that EPA should consider, in the context of the guidance we recommended that EPA develop, whether it needs to reevaluate any of its past determinations to not regulate in light of the systemic and individual shortcomings we have identified in this report.

In response to our recommendation regarding the independent review of the recommended guidance—which EPA has said is not "practicable" to develop—EPA states that it will "plan to seek input" from its National Drinking Water Advisory Council on how to improve transparency and clarity of its regulatory determinations "at an appropriate stage of the process during the agency's (next or future) regulatory determinations effort." We continue to believe it is essential for EPA to develop the guidance we recommend and that the most efficient and effective way to obtain input from one of EPA's drinking water advisory committees is to have it review EPA's draft guidance. Regarding the recommendation for EPA to develop and implement an internal review mechanism to help ensure that EPA's regulatory determinations are consistent with the guidance we recommend, EPA states that it will continue to use its Action Development Process and agency work group to ensure that determinations are clear, transparent, and "as consistent as possible." Our report shows that these mechanisms have not been effective to date. Consequently, we continue to believe that EPA needs to develop and implement an internal review mechanism to help ensure that the agency's regulatory determinations are consistent with the guidance we recommend.

Finally, in response to our recommendation that the EPA Administrator require the Office of Water to include in the public record communications with OMB and other federal agencies during the development of regulatory determinations and associated notices and scientific analyses, EPA states that Executive Order 12866 does not require that every communication with OMB or other federal agencies be included in the public record. EPA also states that unless otherwise required by law, it does not believe that including these documents in the docket is a good policy because the predecisional documents may be confusing to the public, undermine the ultimate policy choice, and inhibit deliberations. We continue to believe that to improve transparency of these determinations, which are by law committed to the Administrator's judgment, EPA should consistently provide in the public record documentation of OMB's and other federal agencies' comments on and revisions to EPA regulatory determination documents, regardless of whether there is a specific legal requirement for disclosure. As a result of EPA's comments on the scope of our recommendation, we have revised it to focus on the transparency of OMB's and other federal agencies' comments on and revisions to EPA's regulatory determination notices and support documents. EPA also provided technical comments, which we incorporated as appropriate.

David C. Trimble
Acting Director
Natural Resources and Environment

APPENDIX I: OBJECTIVES, SCOPE, AND METHODOLOGY

Our objectives were to (1) evaluate the extent to which the Environmental Protection Agency's (EPA) implementation of the 1996 amendments' requirement to determine which potentially harmful contaminants to regulate has helped assure the public of safe drinking water and (2) review the process and scientific analyses EPA used to develop its 2008 preliminary regulatory determination on perchlorate.

To assess EPA's implementation of the 1996 amendments' requirement to determine which potentially harmful drinking water contaminants to regulate, we reviewed the statute, legislative history, and relevant amendments and analyzed relevant documentation, such as *Federal Register* notices announcing EPA's regulatory determinations, contaminant candidate lists, and unregulated contaminant monitoring rules; EPA regulatory determination work group documents; regulatory determination support documents[1]; and relevant policy and guidance documents such as EPA's policy regarding the estimation of relative source contributions, as well as the agency's guidance for considering children's health when developing EPA actions. We reviewed the primary support documents for the 20 contaminants with final regulatory determinations from the 2003 and 2008 cycles and related drinking water health advisories; reports by the Department of Health and Human Services' (HHS) Agency for Toxic Substances and Disease Registry (ATSDR) and the National Academies; and Integrated Risk Information System (IRIS) assessments, as applicable. We reviewed EPA's information on the potential adverse health effects of, and extent to which the public may be exposed to, individual contaminants in public drinking water systems, as well as public comments on EPA's determinations. We judgmentally selected 6 of the 20 contaminants with completed regulatory determinations (boron; dieldrin; 2,4-dinitrotoluene; 2,6-dinitrotoluene; manganese; and 1,1,2,2-tetrachloroethane) to review how EPA presented its rationale and support for its determinations in its regulatory determination notices and key support documents. We selected contaminants that, taken together, provided coverage for the following five variables: (1) contaminants from each of the regulatory determination cycles; (2) contaminants whose occurrence data came from one of the three primary data sources (National Inorganic and Radionuclide Survey, Unregulated Contaminant Monitoring Rounds 1 and 2, and EPA's first Unregulated Contaminants Monitoring Rule); (3) contaminants with cancer adverse health effects as well as contaminants with noncancer adverse health effects; (4) contaminants whose occurrence data had minimum reporting levels higher than the health reference levels, as well as those whose occurrence data had minimum reporting levels lower than the health reference levels; and (5) contaminants detected at levels in excess of the health reference level at 1 percent or more of public water systems sampled, as well as contaminants with fewer such detections.

In addition, we reviewed health effects information and public drinking and source water occurrence data from other entities, such as ATSDR, California's Office of Environmental Health Hazard Assessment, and the U.S. Geological Survey (USGS). We interviewed officials from EPA's Office of Water, Office of Research and Development, and Office of Children's Health Protection to obtain their perspectives on, among other things, the regulatory determination process and the health effects and contaminant occurrence research used to support determinations. We interviewed officials at the Association of State Drinking Water Administrators and the American Water Works Association to obtain their perspectives on EPA's regulatory determinations process and decisions. Overall, the focus of this review has been on EPA's regulatory determination process. While this work included reviewing information developed under two other processes integral to the regulatory determination process—the development of contaminant candidate lists and implementation of the unregulated contaminants monitoring rule—we did not evaluate EPA's implementation of these processes.

To review the process and scientific analyses EPA used to develop its 2008 preliminary regulatory determination on perchlorate, we analyzed the 2008 *Federal Register* notice announcing EPA's preliminary determination and the 2009 *Federal Register* notice—EPA's supplemental request for comments. We reviewed the process EPA used in adopting the National Academies' risk assessment of perchlorate in its IRIS database, as well as other key scientific analyses EPA relied on in making its preliminary determination: its analysis of exposure to perchlorate from food and water and its sensitivity analysis of various age groups to perchlorate exposure. We analyzed the related peer reviews on these analyses and public comments on EPA's preliminary regulatory determination. In addition, we reviewed relevant documentation of communications among EPA officials and between EPA officials and officials from other federal agencies related to perchlorate. Further, we interviewed key officials in EPA's Office of Water and Office of Research and Development who were involved in the development of EPA's preliminary determination. In addition, we interviewed officials from EPA's Office of Water, Office of Research and Development, and Office of Children's Health Protection to obtain information on EPA's decision making, support documents, and analyses regarding the perchlorate decision, as well as to obtain their perspectives on the determination.

In addition, to provide supplemental information on EPA's implementation of the Safe Drinking Water Act of interest to the requesters, we reviewed EPA's regulatory actions from 1974 to 2010 that occurred in response to the major amendments and provide this information in the background section of the report, with additional details provided in appendix II. We also reviewed a key EPA Report to Congress from 1993, which discussed the implementation of the act under the first two legislative frameworks, as well as the conditions preceding the 1996 amendments to the act. We visited a drinking water treatment facility and laboratory to obtain a contextual understanding of public drinking water systems and contaminant occurrence testing, as well as the system operators' perspectives on emerging contaminants and EPA's regulatory determinations process. In addition, we attended the Association of State Drinking Water Administrators' Annual Conference in October 2009 and the American Water Works Association's Water Quality Technology Conference in November 2009 to obtain perspectives on drinking water treatment and regulations, as well as to familiarize ourselves with emerging contaminants of concern and other water quality issues. We conducted this performance audit from March 2009 to May 2011, in accordance with generally accepted government auditing standards. Those standards require that we plan and perform the audit to obtain sufficient, appropriate evidence to provide a reasonable basis for our findings and conclusions based on our audit objectives. We believe that the evidence obtained provides a reasonable basis for our findings and conclusions based on our audit objectives.

Appendix II: Information on EPA's Regulatory Actions under the 1974 Safe Drinking Water Act

EPA's regulatory actions under the Safe Drinking Water Act have varied over time as the agency's legal authority to determine which drinking water contaminants to regulate, if any, has changed from discretionary to prescriptive and back to largely discretionary.

The Act's Original Framework Gave EPA Broad Discretion in Deciding Whether to Regulate Specific Contaminants

In light of increased awareness of widespread water quality problems and health risks, Congress passed the Safe Drinking Water Act in 1974. The act directed the EPA Administrator to develop regulations for drinking water contaminants that "in his judgment" may have adverse human health effects and provided substantial discretionary authority to carry out this mission by not legally mandating the regulation of any specific contaminants or class of contaminants. EPA interpreted the law as providing the agency with discretion to select contaminants for regulation on the basis of their potential for causing an adverse health effect.[1] In contrast to current statutory requirements, the original act did not require EPA to demonstrate the presence of contaminants with potential adverse health effects in public drinking water systems or evaluate public exposure to them. From 1974 to 1986, the act outlined a three-step framework for developing regulations, starting with quickly promulgating interim regulations that would be subject to future revision as national primary drinking water regulations. In addition, EPA was to arrange for a study by the National Academies[2] to assess the health effects of contaminants in drinking water and to develop proposals for maximum contaminant level goals[3]—levels at which there are no known or anticipated adverse human health effects—for those contaminants that may cause adverse health effects. Finally, on the basis of the National Academies' study, EPA was to propose and promulgate maximum contaminant level goals and enforceable maximum contaminant levels that were to be set as close as feasible to the goals for the contaminants that EPA determined may have adverse health effects.

Under this original framework, from 1974 until the 1986 amendments to the act, EPA established interim regulations for 23 contaminants, most of which stemmed from previously established Public Health Service standards.[4] EPA did not, however, propose or promulgate national primary drinking water regulations, referred to as revised regulations, before Congress amended the act in 1986. The National Academies issued reports in 1977, 1980, and 1982 addressing drinking water contaminants of concern; however, the reports did not provide the required maximum contaminant level goals for these contaminants, noting that the development of such goals represented a regulatory function more appropriate for EPA. In 1982 and 1983, EPA issued two Advanced Notices of Proposed Rulemaking based on the National Academies' reports and other scientific sources that together identified 83 contaminants the agency planned to evaluate for regulation; the 83 contaminants included 22 contaminants for which EPA had set interim regulations and 61 additional contaminants.[5] EPA requested public comments on the appropriateness of regulating the identified contaminants, as well as on the agency's proposed regulatory approach. By 1986, when the act was amended, EPA had not completed evaluating the 83 contaminants.

The 1986 Amendments Mandated the Regulation of 83 Specific Contaminants, Most of Which EPA Accomplished, but Not within the Prescribed Time Frames

According to a key EPA report, Congress enacted major amendments to the Safe Drinking Water Act in 1986 because of continuing concerns about drinking water quality and frustration with the pace at which the agency was working to develop drinking water regulations.[6] Importantly, the amendments mandated that by June 1989 EPA regulate the 83 contaminants the agency had previously identified. The amendments also required EPA to regulate a minimum of 25 additional contaminants every 3 years after 1989, known as the "25 in 3" requirement. By directing EPA to regulate 83 specific contaminants and requiring a minimum number of additional regulations every 3 years, Congress curtailed the broad discretion that the 1974 act had initially provided to EPA.

EPA struggled to promulgate the regulations mandated by the 1986 amendments within the prescribed time frames. While EPA did not meet the 1989 deadline, by 1996 the agency had regulated 77 of the mandated contaminants.[7] EPA regulated 6 more of the mandated contaminants by 2001; EPA has not regulated 2 of the 83—radon and sulfate.[8] Further, by 1996, EPA had not regulated any additional contaminants as part of the "25 in 3" requirement but had regulated 1 additional contaminant not statutorily required[9] and had initiated regulations for 8 additional contaminants, which were finalized in 1998 and 2000.

In addition to the requirements to regulate certain contaminants, the 1986 amendments directed EPA to establish a program to test for unregulated contaminants. Specifically, pursuant to the amendments, EPA was required to promulgate regulations requiring every public water system to conduct testing for unregulated contaminants at least once every 5 years.[10] The regulations were to list the unregulated contaminants for which public water systems may be required to test and include criteria by which states could show cause for addition or deletion of contaminants from the list based on an assessment of the contaminants' potential to be found in the system. Accordingly, EPA required two cycles of public water system testing from 1988 to 1992 and from 1993 to 1997. Through this testing, EPA obtained occurrence data—information about where, how frequently, and at what concentrations contaminants are found—on 62 unregulated contaminants from 40 states in the first round and on 48 unregulated contaminants from 35 states in the second round.

By the early 1990s, EPA, states, and drinking water officials had become concerned that continuing the rapid regulatory pace required by the 1986 amendments might overburden drinking water systems and states and hinder effective implementation of the regulations. Consequently, EPA and others called for reforms to better enable the agency to focus on contaminants of greatest public health concern.

The 1996 Amendments Largely Restored EPA's Discretion in Deciding Whether to Regulate Specific Contaminants, and EPA Selected No Contaminants for Regulation until February 2011, When It Determined It Will Regulate Perchlorate

In 1996, in the midst of concerns about the number and pace of regulations required under the 1986 amendments and the need to better focus limited resources on contaminants of greatest public health concern, Congress again enacted significant amendments to the Safe Drinking Water Act, largely restoring EPA's regulatory discretion. In the 1996 amendments, Congress removed the statutory requirement that EPA regulate 25 new contaminants every 3 years; rather, the agency is now required only to decide *whether or not to regulate* at least 5 contaminants every 5 years— called the regulatory determination process. In making regulatory determinations, EPA is to consider contaminants that present "the greatest public health concern," taking into account sensitive populations, among other factors. EPA's regulatory determinations are to be based on three broad statutory criteria, all of which must be met for EPA to decide that a regulation is needed:

- the contaminant may have an adverse effect on the health of persons;
- the contaminant is known to occur or there is a substantial likelihood that the contaminant will occur in public water systems with a frequency and at levels of public health concern; and
- in the sole judgment of the Administrator, regulation of such contaminant presents a meaningful opportunity for health risk reduction for persons served by public water systems.

While the 1996 amendments largely restored discretion in EPA's regulatory authority, they also specified two other requirements related to identification and evaluation of contaminants that present the greatest public health concern. First, the act established a new requirement for EPA to identify and publish every 5 years a list of unregulated contaminants that may pose risks through drinking water, called the contaminant candidate list, from which EPA is to select contaminants for regulatory determinations.[11] Second, the amendments modified the authority EPA had under the 1986 amendments to require public water systems to test for unregulated contaminants. Specifically, the 1996 amendments limited both the number of small public water systems that EPA could require to conduct testing and the number of unregulated contaminants for which systems must test. While EPA retained its authority to require that all large public water systems (those serving more than 10,000 people) test for certain unregulated contaminants, it may now only require a representative sample of small public water systems (those serving 10,000 or fewer people) to test for those contaminants. In addition, whereas the 1986 amendments placed no limit on the number of unregulated contaminants that EPA could require public water systems to test, the 1996 amendments limited EPA to requiring testing for a maximum of 30 unregulated contaminants every 5 years. EPA has implemented these requirements through its testing program for unregulated contaminants. In contrast to the testing previously required by EPA and conducted by public water systems pursuant to the 1986 amendments, which relied on varying protocols, EPA's implementation of the testing requirements under the 1996 amendments uses standardized testing protocols it has developed to help ensure more consistent data quality and comparability.

Under the 1996 amendments, if EPA decides through its regulatory determination process that a drinking water regulation is warranted, the agency has 24 months to publish a proposed rule and an additional 18 to 27 months to issue a final drinking water rule setting an enforceable limit for the contaminant—the maximum contaminant level. With regard to developing maximum contaminant levels, the 1996 amendments added a requirement that EPA conduct a cost-benefit analysis as part of the standard-setting process. The amendments further require that in carrying out the provisions concerning listing, selecting, and regulating contaminants, to the degree that an action is based on science, EPA use the best available, peer-reviewed science and data collected by accepted or best available methods. EPA is also to ensure that the presentation of information on public health effects is comprehensive, informative, and understandable.

In addressing the requirement to make decisions on whether to regulate at least 5 contaminants every 5 years, EPA made regulatory determinations on 20 contaminants in 2003 and 2008, deciding in each case not to regulate. EPA had not decided to regulate any additional contaminants since the enactment of the 1996 amendments[12] until February 2011 when the EPA Administrator announced that the agency made a determination to regulate perchlorate as well as a group of carcinogenic volatile organic compounds (VOC), which include chemicals such as certain industrial solvents. This planned regulatory action stems from EPA's effort to revise the existing drinking water regulations for trichloroethylene (TCE) and tetrachloroethylene (PCE), and the Administrator's plan to address contaminants as groups rather than one at a time to enhance drinking water protection in a timely and cost-effective manner. According to EPA, the agency plans to include 8 regulated contaminants, as well as 8 unregulated VOCs that are on its current candidate list.

Regarding the requirement to develop, every 5 years, a contaminant candidate list, EPA issued such lists in 1998, 2005, and 2009. Pursuant to its authority to require testing of unregulated contaminants, EPA mandated one cycle of testing, which was completed in 2005; a second cycle is expected to be completed in mid-2011, according to EPA officials.

APPENDIX III: INFORMATION ON U.S. PUBLIC DRINKING WATER SYSTEMS

The following tables provide detailed information from the U.S. Environmental Protection Agency (EPA) on public drinking water systems in the United States. EPA's Office of Water Web site defines three types of public water systems:

- Community Water System: a public water system that supplies water to the same population year-round.
- Non-Transient Non-Community Water System: a public water system that regularly supplies water to at least 25 of the same people at least 6 months per year, but not year-round. Some examples are schools, factories, office buildings, and hospitals that have their own water systems.
- Transient Non-Community Water System: a public water system that provides water in a place such as a gas station or campground where people do not remain for long periods of time and is open at least 60 days a year.

Table 6 provides information on the number of systems and population served, including percentages, by system type and size. Table 7 provides information on the number of systems and population served, including percentages, by system type and source water.

Table 6. U.S. Public Drinking Water Systems by Size

Category of system	Number of systems	Percentage of systems	Population served	Percentage of population served
Community water systems				
Very small systems[a]	28,804	56	4,820,949	2

Category of system	Number of systems	Percentage of systems	Population served	Percentage of population served
Small systems[b]	13,820	27	19,806,741	7
Medium systems[c]	4,871	9	28,402,697	10
Large systems[d]	3,746	7	106,856,965	36
Very large systems[e]	410	1	134,452,529	46
Total	51,651	100	294,339,881	100
Non-transient non-community water systems				
Very small systems	15,619	85	2,195,162	35
Small systems	2,625	14	2,704,116	43
Medium systems	132	1	699,947	11
Large systems	18	<1	440,980	7
Very large systems	1	<1	203,000	3
Total	18,395	100	6,243,205	100
Transient non-community water systems				
Very small systems	80,703	97	7,147,163	54
Small systems	2,681	3	2,598,895	20
Medium systems	87	<1	471,533	4
Large systems	11	<1	360,715	3
Very large systems	2	<1	2,725,000	20
Total	83,484	100	13,303,306	100
All public water systems				
Very small systems	125,126	82	N/A[f]	N/A[f]
Small systems	19,126	12	N/A[f]	N/A[f]
Medium systems	5,090	3	N/A[f]	N/A[f]
Large systems	3,775	2	N/A[f]	N/A[f]
Very large systems	413	>1	N/A[f]	N/A[f]
Total	153,530	100	N/A[f]	N/A[f]

Source: EPA.

[a] Very small water systems serve 25–500 people.

[b] Small water systems serve 501–3,300 people.

[c] Medium water systems serve 3,301–10,000 people.

[d] Large water systems serve 10,001–100,000 people.

[e] Very large water systems serve more than 100,000 people.

[f] Populations are not summed because some people are served by multiple systems and counted more than once.

Table 7. U.S. Public Drinking Water Systems by Source Water

Category of system	Number of systems	Percentage of systems	Population served	Percentage of population
Community water systems				
Groundwater[a]	40,025	78	88,032,021	30
Surface water[b]	11,617	22	206,264,490	70
Total	51,642	100	294,296,511	100
Non-transient non-community water systems				
Groundwater	17,688	96	5,415,937	87

Table 7. (Continued)

Category of system	Number of systems	Percentage of systems	Population served	Percentage of population
Surface water	702	4	820,476	13
Total	18,390	100	6,236,413	100
Transient non-community water systems				
Groundwater	81,492	98	10,754,201	81
Surface water	1,978	2	2,548,200	19
Total	83,470	100	13,302,401	100
All categories of systems				
Groundwater	139,205	91	N/A[c]	N/A[c]
Surface water	14,297	9	N/A[c]	N/A[c]
Total	153,502	100	N/A[c]	N/A[c]

Source: EPA.

[a]Groundwater: Groundwater comes from natural underground formations, often consisting of sand or gravel, that contain water. These formations are called aquifers.

[b]Surface water: Surface water sources include lakes, streams, rivers, and reservoirs. Groundwater sources under the direct influence of surface water are also included in this category.

[c]Populations are not summed because some people are served by multiple systems and counted more than once.

APPENDIX IV: CALCULATIONS EPA USES TO DEVELOP HEALTH REFERENCE LEVELS FOR DRINKING WATER CONTAMINANTS BEING CONSIDERED FOR REGULATION

This appendix provides information on the method EPA generally has used to compute the health reference level for drinking water contaminants that are carcinogenic and the method it uses for contaminants with noncarcinogenic adverse health effects, such as neurological disorders.

$$\text{Health reference level} = \text{Concentration in drinking water equivalent to one-in-a-million } (10^{-6}) \text{ cancer risk} = \frac{(\text{Risk}^{a} \times \text{body weight}^{b})}{(\text{Slope factor}^{c} \times \text{drinking water intake}^{d})}$$

Source: EPA.

[a]Expression of increased cancer risk from a lifetime of exposure: 1 person in a million = 10^{-6} = 0.000001.

[b]Adult body weight, assumed to be 70 kilograms.

[c]Slope factors result from EPA modeling of linear low-dose extrapolations. A slope factor is an upper bound, approximating a 95 percent confidence limit, on the increased cancer risk from a lifetime exposure to an agent by ingestion. This estimate is usually expressed in units of proportion (of a population) affected per milligram of substance per kilogram of body weight per day (mg/kg/day).

[d]Amount of water consumed by an adult per day, assumed to be 2 liters.

Figure 1. Health Reference Level Equation for Contaminants with Carcinogenic Health Effects.

$$\text{Health reference level} = \underbrace{\left[\frac{\text{(Reference dose}^{a} \times \text{body weight}^{b}\text{)}}{\text{drinking water intake}^{c}}\right]}_{\text{Drinking water equivalent level}^{d}} \times \text{Relative source contribution}^{e}$$

Source: EPA.

[a]An estimate of a daily exposure to the human population (including sensitive subpopulations) that is likely to be without an appreciable risk of deleterious effects during a lifetime, generally expressed in units of milligrams per kilogram of body weight per day (mg/kg/day).

[b]Adult body weight, assumed to be 70 kilograms.

[c]Amount of water consumed by an adult per day, assumed to be 2 liters.

[d]Estimated exposure to a contaminant that is assumed to be protective for noncarcinogenic health effects during a lifetime of exposure, generally expressed in units of milligrams per liter (mg/L). It is calculated by multiplying the reference dose times body weight and dividing that result by drinking water intake.

[e]The estimate of the exposure to a contaminant from drinking water relative to overall exposure from other sources (e.g., food and ambient air), expressed as a percentage.

Figure 2. Health Reference Level Equation for Contaminants with Noncarcinogenic Adverse Health Effects.

APPENDIX V: CALCULATIONS OF RELATIVE SOURCE CONTRIBUTION USING THE PERCENTAGE AND SUBTRACTION METHODS

As discussed in the report, to determine a health reference level for a contaminant with adverse noncarcinogenic health effects for use in its regulatory determinations, EPA typically applies a relative source contribution factor—the allocation of the oral exposure to the contaminant from drinking water alone—in its equation (see fig. 3).

$$\text{Health reference level} = \underbrace{\left[\frac{\text{(Reference dose}^{a} \times \text{body weight}^{b}\text{)}}{\text{drinking water intake}^{c}}\right]}_{\text{Drinking water equivalent level}^{d}} \times \text{Relative source contribution}^{e}$$

Source: EPA.

[a]An estimate of a daily exposure to the human population (including sensitive subpopulations) that is likely to be without an appreciable risk of deleterious effects during a lifetime, generally expressed in units of milligrams per kilogram of body weight per day (mg/kg/day).

[b]Adult body weight, assumed to be 70 kilograms.

[c]Amount of water consumed by an adult per day, assumed to be 2 liters.

[d]Estimated exposure to a contaminant that is assumed to be protective for noncarcinogenic health effects during a lifetime of exposure, generally expressed in units of milligrams per liter (mg/L). It is calculated by multiplying the reference dose times body weight and dividing that result by drinking water intake.

[e]The estimate of the exposure to a contaminant from drinking water relative to overall exposure from other sources (e.g., food and ambient air), expressed as a percentage.

Figure 3. Health Reference Level Equation.

According to EPA policy, if the agency determines that it has adequate data to estimate the relative source contribution—instead of applying a default assumption—to calculate a health reference level, it may use one of two methods—the percentage or subtraction method.[1] Tables 9 and 10 illustrate the differences in the health reference levels that result when each of these methods is applied using the hypothetical exposure data and reference dose in table 8.

Table 8. Hypothetical Data Used in Examples

Estimated exposure from water (E_W) =	0.15	µg/kg/day
Estimated exposure from food (E_F) =	0.15	µg/kg/day
Total estimated exposure from water and food (E_W+E_F=E_{TOT}) =	0.3	µg/kg/day
Reference dose (RfD) =	0.5	µg/kg/day

Source: GAO.

Example 1: Percentage Method

The percentage method is a comparison of multiple sources of exposure with one another to estimate their relative contribution to the total. It is intended to reflect the exposure to a contaminant from drinking water relative to other sources of exposure, as well as the likelihood for ever-changing levels in each of the sources of exposure (due to ever-changing sources of emissions and discharges). Importantly, it is based on an assumption that there may be enough relative variability in exposure such that an apportionment (relating that percentage to the RfD) is a reasonable way of accounting for the uncertainty regarding that variability. Using the percentage method, the relative source contribution is determined by (1) calculating the relative proportion of exposure from water as a percent of the total observed exposure and then (2) applying that percentage as the relative source contribution in the health reference level equation. See example in table 9.

Table 9. Example of How the Relative Source Contribution Is Determined Using the Percentage Method

Step 1: Calculate the relative proportion of exposure from water as a percent of the total observed exposure.				
Estimated exposure from water (E_W)	÷	Total estimated exposure (E_{TOT})	=	Relative source contribution
0.15 µg/kg/day	÷	0.3 µg/kg/day	=	0.5 (50 percent)

Step 2: Apply relative source contribution percentage in health reference level equatio								
Health reference level (parts per billion)	=	[(Reference dose (µg/kg/day)	x	Body weight) (kilograms)	÷	Drinking water intake] (liters per day)	x	Relative sourcecontribution (expressed as apercentage)
8.8	=	[(0.5	x	70)	÷	2]	x	50

Source: GAO.

Example 2: Subtraction Method

The subtraction method allocates the entire reference dose to the known sources of exposure by subtracting the known nontarget sources of exposure and allocating the remainder of the reference dose to the target—in this case, drinking water—even in cases (such as this example) where the total estimated exposure is less than the reference dose. This method has the effect of removing any cushion between the existing exposure levels and the reference dose. Therefore, using this method may allow drinking water exposures in excess of levels currently found in public water systems. To calculate the relative source contribution using the subtraction method: (1) subtract all non-drinking-water exposures from the reference dose to determine the amount of the reference dose "available" for exposure through drinking water, (2) determine what percentage of the reference dose that remainder represents, and (3) apply the resulting percentage as the relative source contribution in the health reference level equation. See example in table 10.

Table 10. Example of How the Relative Source Contribution Is Determined Using the Subtraction Method

Step 1: Subtract nonwater exposures from reference dose.				
Reference dose (RfD)	–	Estimated exposure from food (E_F)	=	Remainder of reference dose "available" for water
0.5 µg/kg/day	–	0.15 µg/kg/day	=	0.35 µg/kg/day

Step 2: Determine the amount of the reference dose "available" for exposure from drinking water by dividing the result from Step 1 by the reference dose and multiplying by 100 to convert to a percentage.				
Remainder "available" for water	÷	Reference dose (RfD)	=	Relative source contribution
0.35 µg/kg/day	÷	0.5 µg/kg/day	=	0.7 (70 percent)

Step 3: Apply the relative source contribution percentage in the health reference level equation.								
Health reference level (parts per billion)	=	[(Reference dose (µg/kg/day)	x	Body weight) (kilograms)	÷	Drinking water intake] (liters per day)	x	Relative sourcecontribution (expressed as apercentage)
12.3	=	[(0.5	x	70)	÷	2]	x	70

Source: GAO.

APPENDIX VI: SUPPLEMENTAL INFORMATION ON EPA'S 2003 REGULATORY DETERMINATION FOR MANGANESE AND ITS 2008 DETERMINATION FOR BORON

This appendix provides supplemental information on the agency's use of outdated and limited occurrence data and inconsistent consideration and presentation of potential health risks in its regulatory determination notices and support documents for manganese and boron.

EPA's 2003 Regulatory Determination to Not Regulate Manganese

Occurrence Data

A ubiquitous, naturally occurring element found in certain rocks and sediments, manganese is also produced and used in a wide variety of industrial processes and consumer products, resulting in widespread manmade releases of manganese and manganese compounds into the environment.[1] EPA made its decision in 2003 to not regulate manganese using data on the occurrence and likely occurrence of manganese in drinking water that the agency acknowledged was outdated and limited. Specifically, EPA relied primarily on older (1980s), but nationally representative, data from its National Inorganic and Radionuclide Survey to assess manganese occurrence in public water systems fed by groundwater sources.[2] In its regulatory determination support document for manganese, EPA reported that 3.2 percent of groundwater public drinking water systems exceeded EPA's health reference level for manganese—affecting an estimated 2.3 million people. In addition, EPA reported that 4.6 percent of groundwater systems exceeded one-half the health reference level—affecting an estimated 3.9 million people. In its *Health Effects Support Document for Manganese*, EPA stated that it "should be noted that these estimates are based on very limited and outdated data. The possibility exists that the number of people served by groundwater with manganese levels that are above the health reference level could be higher than these estimates; however, the data are lacking at this time to develop a more timely assessment."[3] Moreover, the use of older data is problematic because of significant increases in releases of manganese and manganese compounds to the environment; according to EPA's 2003 regulatory determination support documentation, the increases since the 1990s have been dramatic in some years. Despite EPA's acknowledgement of significant deficiencies in the underlying data, and its decision not to update these data through its testing program, EPA's notice did not explain why it nonetheless proceeded with a regulatory determination for manganese. Further, as discussed in the body of this report, EPA has not issued guidance on what threshold levels of occurrence would satisfy the second statutory criterion for a decision to regulate. In this case, EPA decided that levels of manganese in excess of the health reference level in at least 3.2 percent of groundwater drinking water systems did not warrant regulation.[4]

Moreover, EPA did not have national data on manganese occurrence in public water systems that use surface water—according to EPA, surface water systems serve about 70 percent of community water system customers. While EPA limited its occurrence estimates to groundwater systems in its 2003 manganese regulatory determination notice, its support document for manganese discusses some USGS data on ground and surface source water and manganese occurrence data from five states, but in a manner that does not explain the determination in light of this potentially inconsistent data.5, [6] For example, the agency characterized drinking water occurrence data on manganese from five states as showing "substantial low-level manganese occurrence," but did not reconcile this assessment with the occurrence data it presented for these five states that show the following percentages of state populations served by public water systems with levels of manganese that exceed the health reference level: 2.4 percent, 3.2 percent, 9.1 percent, 14.7 percent, and 27.2 percent.[7] In addition, EPA described USGS detections in excess of its health reference level at groundwater sites (14 percent of samples) and surface water sites (10 percent of samples) as "modest" occurrence and "relatively low" occurrence, respectively.

Further, EPA did not include information about the large number of Superfund sites known to be contaminated with manganese in its regulatory determination documents. According to ATSDR, manganese has been found in at least 51 percent of EPA's 1,699 Superfund (National Priorities List) sites, which include some of the most seriously contaminated hazardous waste sites in the nation.[8] ATSDR's *Public Health Statement for Manganese* notes that this information is important because the manganese at these sites may be harmful to the public. Moreover, EPA's regulatory determination documents do not mention that DOD's emerging contaminants program has been monitoring manganese. Under this program, a DOD team on "materials of evolving regulatory interest" is focusing on a number of contaminants, including manganese, to develop actions to respond to such potential factors as health impacts, cleanup costs, compliance costs, readiness impacts, and facilities life-cycle costs.[9]

Health Effects Data

Relying on the third statutory criterion of presenting a meaningful opportunity for health risk reduction for this determination, EPA concluded that because manganese is generally not considered to be very toxic when ingested with the diet (i.e., food), and since drinking water accounts for a relatively small proportion of manganese intake, regulation would not likely present a meaningful opportunity for health risk reduction for persons served by public drinking water systems. However, EPA does not explain how this determination relates to information EPA had indicating that manganese in drinking water could have significant adverse health effects on sensitive populations, including children. Specifically, EPA's 1996 IRIS assessment identified children, pregnant women, elderly people, iron- or calcium-deficient individuals, and individuals with liver impairment as potentially sensitive populations. Individuals in these populations may have an increased potential for excessive amounts of manganese in the body because of increased absorption or altered clearance mechanisms.

EPA's 1996 IRIS assessment notes that differences in absorption and clearance mechanisms may be of particular importance for those exposed to manganese by multiple routes. Regarding oral exposure, the IRIS assessment states that there is some evidence that infants absorb more manganese from the gastrointestinal tract than adults, that newborns are less able to excrete absorbed manganese, and that the absorbed manganese more easily passes the blood-brain barrier in infants.[10] EPA's IRIS assessment also states that there is a concern for infants fed formula because it typically has a much higher concentration of manganese than human milk, and if powdered formula is made with drinking water, any manganese in the water would represent an additional source of intake.[11] The IRIS assessment states that these considerations concerning increased exposure in an important population group "warrant caution until more definitive data are available," noting in addition the likelihood that any adverse neurological effects of manganese are likely to be irreversible and not manifested for many years after exposure.

The 1996 IRIS assessment cites scientific literature up to 1994; since then, additional studies have identified health risks for children, particularly infants, from manganese-contaminated water. For example, a recent risk assessment journal article on the health risks to children from exposure to manganese in drinking water highlighted four epidemiological studies and two case reports published between 1994 and 2007 that demonstrate a plausible association between elevated manganese concentrations in drinking water and altered neurological function in children.[12] Effects that were identified include reduced intellectual function and lower performance scores on behavioral tests.[13] In addition, a 2006 report by the National Academies, *Spacecraft Water Exposure Guidelines for Selected Contaminants,* stated that a survey of the literature on manganese toxicity strongly indicates, among other things, "hypersusceptibility" of infants and newborns and the elderly. Moreover, on the basis of information obtained from literature reviews conducted between 2001 and 2003 on chemicals with existing assessments to determine if new scientific information might change a current IRIS assessment, the IRIS program decided that manganese should be reassessed.[14] Studies continue to identify children as particularly sensitive to manganese contamination in drinking water.[15]

Although the IRIS assessment had identified newborns as a potential sensitive subpopulation that warranted caution until more definitive data became available—and new data continued to identify concerns for infants and children—the Office of Water did not develop a specific health reference level for infants or children and did not make other adjustments to its health reference level to assess the sensitivity of children to manganese in drinking water. For example, Office of Water officials told us that one way it can assess sensitive populations is by adjusting the health reference level downward to account for increased sensitivity or by developing a separate risk assessment for the sensitive population. However, the Office of Water derived its drinking water health reference level—0.3 milligrams of manganese per liter of water—from the IRIS reference dose without any additional adjustment for sensitive populations beyond those factored into the reference dose. Specifically, the health reference level was calculated using EPA's 1996 reference dose[16] in its standard formula to develop a drinking water equivalent level, using the average weight and daily water consumption of a healthy adult.[17]

Moreover, EPA discussed the health risks to infants in an inconsistent and, at times, incomplete manner in its regulatory determination documents. For example, EPA's 2003 regulatory determination support document for manganese states unequivocally that there are "no data to indicate children are more sensitive to manganese than adults." However, EPA's 2003 health effects support document for manganese discusses studies that identify an association between exposure to manganese in drinking water and learning disabilities in children and concludes that additional studies are needed to investigate the possibility that children are more sensitive than adults. In addition, while EPA's regulatory determination support document for manganese notes that infants and newborns may be potentially susceptible to manganese toxicity, this key document does not disclose that newborns may be exposed to high levels of manganese from infant formula or that these high levels of manganese in formula can be magnified when it is reconstituted with manganese-contaminated water. The support document contains only the following statement on manganese in breast milk and infant formula: "Although the manganese content in a soy-based formula is higher than the manganese content in human milk, the actual absorption of manganese from the formula may not be substantially greater, since soy milk is high in phytate[18] and vegetable protein." Importantly, EPA's health effects support document also contains the prior statement but then specifically cites the results of several studies that "argue against this possibility"—that is, the cited studies contradict the hypothesis presented in the regulatory determination support document. Further, while EPA's health effects document discloses that infant formula typically contains a much higher concentration of manganese than human or cows' milk and that powdered formula reconstituted with drinking water represents an additional source of manganese intake for a potentially sensitive population, its regulatory determination support document omits this important exposure information.

Overall, EPA presented the health risks of exposure to manganese in a manner that downplayed the potential risks in part by highlighting a concern about manganese deficiency. Specifically, EPA states multiple times in its regulatory determination and support documents that because manganese is an essential nutrient, concern over potentially toxic effects from high oral exposure must be balanced against concern for adverse effects from manganese deficiency. However, according to a National Academies' report,[19] there is no recommended daily allowance for manganese because it is available in numerous food sources to various degrees and "no natural deficiency of manganese in humans has been encountered." Moreover, in the context of deciding whether manganese in drinking water should be regulated, EPA's concern about the potential for adverse effects from manganese deficiency seems to be inconsistent with the fact that EPA already has established a secondary drinking water standard for manganese that addresses aesthetic problems—largely discoloration—of 0.05 milligrams per liter.[20] This recommended standard is 6 times more stringent than the health reference level of 0.3 milligrams per liter that EPA used in its regulatory determination. Thus, the basis for EPA's concern about manganese deficiency stemming from a possible health-based national primary drinking water standard is not clear.

Although EPA concluded that dietary intake of manganese was not very toxic, the agency determined that there was a need to issue a health advisory in conjunction with the regulatory determination on manganese. According to the advisory, it is to provide "guidance to communities that may be exposed to drinking water contaminated with high manganese (Mn) concentrations" and also "provides guidance on the concentrations below which potential health and organoleptic[21] problems would unlikely occur." The advisory provides a 10-day health advisory of 1 milligram of manganese per liter of water for acute exposure for children. However, EPA recommends that manganese exposure for infants younger than 6 months should be limited to the agency's lifetime (chronic) health advisory level of 0.3 milligrams per liter of water—even for acute exposure—"because of the concerns for differences in manganese content in human milk and formula and the possibility of a higher absorption and lower excretion in young infants." As discussed, manganese levels in excess of the health reference level of 0.3 milligrams per liter of water have been detected in a number of public water systems. However, EPA's health advisories do not provide information on which public water systems may contain levels in excess of these standards. While some systems may test manganese voluntarily or to meet state requirements, current manganese levels may not be available for many systems. The potential lack of monitoring data for a particular system, along with the lack of public notification of health advisories by EPA's Office of Water, may make it difficult for consumers to become aware of any related risks.

EPA's 2008 Regulatory Determination to Not Regulate Boron

Occurrence Data

Boron is a naturally occurring element, and both naturally occurring and manmade borate compounds are used in many products, such as glass, ceramics, soaps, fire retardants, pesticides, cosmetics, photographic materials, and high-energy fuels. According to EPA's regulatory determination documents, the potential adverse health affects that may be associated with exposure to boron in drinking water include adverse effects on male reproductive systems.[22]

As it had for manganese 5 years year earlier, for its 2008 regulatory determination for boron, EPA used nationally representative occurrence data from its National Inorganic and Radionuclide Survey (NIRS) of drinking water systems fed by groundwater sources—data that EPA described in 2003 as “outdated and limited.”[23] Importantly, EPA had limited data on occurrence in systems that use surface water, which serve about 70 percent of the population served by community water systems. That is, EPA’s conclusions on the occurrence of boron in public water systems that use surface water relied on a 2004 voluntary, industry-sponsored survey by the American Water Works Research Foundation that provided data on 113 samples of untreated surface water analyzed for boron.[24, 25]

While none of the 113 samples had levels of boron in excess of one-half of EPA’s health reference level, EPA acknowledged the survey was not statistically representative. As a result, these data may not correctly represent the extent or magnitude of boron in surface water. Along these lines, we note that while the 189 utilities participating in the study were located in 41 states, 41 percent of them were located in 3 states— California, Illinois, and Indiana. Also, unlike EPA’s testing program for unregulated contaminants, which requires all public water systems serving more than 10,000 people and a randomly selected sample of public water systems serving 10,000 or fewer people to conduct either two or four tests during a period of 12 consecutive months, water utilities responding to this survey conducted one test per source. According to EPA’s unregulated contaminants testing program documentation, multiple samples during a year are necessary to capture the annual variability in contaminant occurrence to approach an adequate characterization of potential exposure. Further, results were not provided for 16 percent of the samples, and data were not provided on the number of public water systems associated with the 113 tests.

In addition, EPA did not address whether and how anthropogenic (manmade) releases of boron might affect the uncertainty associated with the surface water occurrence data. For example, while EPA’s regulatory determination acknowledges that the manufacture and use of products containing boron compounds add to the release of boron into the environment, and presents information about environmental releases of boron identified in EPA’s Toxics Release Inventory, EPA does not explain in its regulatory determination documents whether these manmade releases are relevant to its regulatory determination. Further, EPA’s regulatory determination support document for boron does not acknowledge that, according to ATSDR, boron contamination is present at 10 percent or more of the nation’s current or former Superfund sites, some of which could be contaminating actual or potential drinking water sites.[26] ATSDR data also show that the average surface water boron concentration in the United States is lower than EPA’s health reference level—0.1 milligrams per liter and 1.4 milligrams per liter, respectively—but that surface water concentrations vary greatly depending on boron content of local geologic formations as well as manmade sources.

On the basis of the dated but nationally representative groundwater data, EPA reported in support documents that approximately 1.7 percent of groundwater public water systems serving about 400,000 people had detections of boron above the health reference level, and approximately 4.3 percent of water systems serving about 2.5 million people had detections in excess of one-half the health reference level. In its regulatory determination notice, however, EPA reported the unweighted sampling data in NIRS, identifying exposure to about 6,400 people above the health reference level and about 42,700 above one-half the health reference level. Considering the data on groundwater sources and the industry research foundation's finding that boron was not detected above one-half the reference level in surface water, EPA provided the following rationale for not regulating boron:

> "Taking this surface water information into account, the agency believes the overall occurrence and exposure from both surface and groundwater systems together is likely to be lower than the values observed for the NIRS groundwater data. Because boron is not likely to occur at levels of concern when considering both surface and groundwater systems, the agency believes that a national public drinking water regulation does not present a meaningful opportunity for health risk reduction."

This rationale appears to downplay the potential health risk to those served by systems with groundwater sources and suggests that the agency uses a national threshold for occurrence at levels of health concern in making its regulatory determinations. Moreover, it is not clear why EPA did not take advantage of its testing program to obtain more current and complete occurrence data before making its regulatory determination.

Health Effects Data

EPA's presentation of the health risks of exposure to boron in its regulatory determination support documents and the health advisory it developed concurrent with its regulatory determination differ both in terms of the potential for adverse health effects and the levels in drinking water identified as generally safe. Regarding the potential health risks, in its regulatory support document, EPA stated that animal studies identify the developing fetus as potentially sensitive to boron and concluded that boron concentrations greater than the health reference level of 1.4 milligrams of boron per liter of water "might" have an effect on prenatal development.[27] In this document, the Office of Water also states that the primary adverse effects identified from studies of animals after chronic exposure to low doses of boron generally involve the testes and the developing fetus. The Office of Water's May 2008 *Drinking Water Health Advisory for Boron* (published just 2 months before the regulatory determination for boron) provides this same information, but states more strongly that there is "compelling evidence" to suggest that the "testicular morphological effects" reported in studies of animals are applicable to children and concluded that exposure to boron between birth and puberty may result in adverse cellular effects that would "affect testicular function." In addition, a third related document—EPA's *Summary Document from the Health Advisory for Boron and Compounds*— provides an important warning regarding infants' exposure to boron in drinking water that is not included in either EPA's drinking water advisory for boron or its regulatory determination support document. Specifically, the summary document states that water containing boron "at levels above the HA [health advisory]" should not be used to prepare food or formula for infants. EPA does not identify which of the exposure duration health advisories it is referring to in this warning.[28]

Despite considering children to be a sensitive population and EPA's children's health policy suggesting that assessments for infants and children be conducted or—if not conducted, an explanation be provided as to why assessments are not warranted—the Office of Water did not include an assessment for children exposed to boron at the health reference level as part of its regulatory determination. Further, as discussed earlier, during the 2000s, EPA developed technical guidance that can assist the Office of Water in assessing the sensitivity of children to drinking water contaminants. EPA officials told us that the Office of Water had not decided how to include the Office of Children's Health Protection guidance in its assessments at the time of the regulatory determination on boron.

In addition, the health reference level that the Office of Water used to assess an adult's risk of daily, chronic exposure to boron in drinking water for its regulatory determination process—1.4 milligrams of boron per liter of water—is widely divergent from the level that the office concurrently developed for the health advisory it prepared in conjunction with its regulatory determination. Specifically, the health advisory's chronic risk level for adults was 5 milligrams of boron per liter of water.[29] The assessments differ primarily because EPA used the percentage method and the standard default assumption in developing the health reference level, whereas it used the subtraction method in developing the health advisory. In discussing these different assessments with us, EPA officials said the health advisory and health effects support document for the regulatory determination were "out of sync" because of lengthy administrative processes. The officials said new information was identified and included in the health advisory that was not available at the time the regulatory determination on boron was made. As a basis for using the subtraction method for its health advisory, EPA's advisory cited its 2000 *Methodology for Deriving Ambient Water Quality Criteria for the Protection of Human Health*. This document cautions that the subtraction method generally results in health exposure levels that are significantly higher than those that would be derived using the percentage method. Further, on the basis of the exposure data EPA presented in its regulatory determination documents, it is not clear that the agency had the adequate exposure data from all sources that the guidance specifies is a prerequisite for using the subtraction method instead of the more conservative default value.

Moreover, EPA's health advisory does not explain why the agency used the subtraction method for the advisory at the same time the agency was using the percentage method for the regulatory determination. The effect of using the subtraction method is significant because EPA's regulatory determination document stated that the highest observed concentration of boron in groundwater was approximately 3.3 milligrams per liter—a level below the health advisory level of 5 milligrams per liter but above the health reference level of 1.4 milligrams.

Further, because boron is not regulated by EPA, and EPA does not currently require public water systems to test for boron, it would be difficult for most people to determine how much boron is present in their drinking water—requisite information for heeding EPA's warning to not use drinking water from some public water systems to prepare infant food or formula. Except in the states that have issued drinking water guidelines for boron, such as the six identified by EPA, individuals would generally have to have their water tested by a laboratory for the presence of boron.

APPENDIX VII: EPA'S EVALUATION OF PERCHLORATE OCCURRENCE AT TWO LEVELS—5 PARTS AND 15 PARTS PER BILLION OF PERCHLORATE IN WATER

As discussed earlier, before EPA made its preliminary regulatory determination for perchlorate on the basis of the health reference level of 15 parts per billion of perchlorate in drinking water (calculated using a relative source contribution of 62 percent), the Office of

Water had calculated a health reference level of 5 parts per billion using the 20 percent default relative source contribution. Table 11 shows the impact of various relative source contribution factors and their related health reference levels on EPA's characterization of the exposure to perchlorate in drinking water at levels of public health concern, based on a similar table the agency included in its preliminary regulatory determination notice. In particular, the table includes the exposure estimates related to the two health reference levels discussed above—the health reference level of 5 parts per billion that EPA initially developed using the 20 percent default relative source contribution and the health reference level of 15 parts per billion that EPA subsequently developed for its preliminary regulatory determination using data from its novel exposure analysis and the subtraction method.

At a health reference level of 5 parts per billion, EPA found that 3.2 percent of public water systems had at least one detection in excess of that level, which EPA estimated could expose 14.6 million people to perchlorate at the level of public health concern. According to some EPA officials, this amount of public exposure to perchlorate at levels above the health reference level would suggest that a regulation was warranted. In contrast, the amount of public exposure to perchlorate at levels above the 15 parts per billion health reference level was significantly lower. Specifically, at a health reference level of 15 parts per billion, less than 1 percent of public water systems had at least one detection above the health reference level, which EPA estimated could expose about 2 million people to perchlorate in drinking water at the level of public health concern.

Table 11. Perchlorate Occurrence and Population Exposure Estimates at Various Potential Health Reference Levels Reported in EPA's Preliminary Regulatory Determination for Perchlorate

Potential health reference level (parts per billion)	Relative source contribution[a] (percentage)	Public water systems with at least one detection greater than the health reference level (percentage)	Population served by public water systems with at least one detection greater than the health reference level (millions of people)
4	16	4.0	16.6
5	20	3.2	14.6[b]
7	29	2.1	7.2
10	41	1.4	5.0
12	49	1.1	3.6
15	62[c]	0.8	2.0[d]
17	69	0.7	1.9
20	82	0.5	1.5
25[e]	100	0.4	1.0

Source: GAO analysis of EPA data.

[a]We calculated the relative source contribution factors using EPA's equation for calculating health reference levels, using EPA's default assumptions for body weight and drinking water intake.

[b]Exposure to perchlorate that EPA estimated at a health reference level of 5 parts per billion that EPA developed using the 20 percent default relative source contribution.

[c]Using EPA's default assumptions for body weight and drinking water intake to calculate the relative source contribution that corresponds with a perchlorate concentration in drinking water of 15 micrograms per liter results in a relative source contribution of 61.2 percent. We report 62 percent

because that is the relative source contribution EPA derived based on the agency's exposure analysis.

[d]Exposure to perchlorate that EPA estimated at a health reference level of 15 parts per billion that EPA developed using data from its novel exposure analysis and the subtraction method to calculate the relative source contribution.

[e]This value reported by EPA represents the drinking water equivalent level of 24.5 parts per billion rounded to the nearest whole number. A health reference level equal to the drinking water equivalent level assumes that all exposure to the contaminant is from drinking water—hence a relative source contribution of 100 percent.

APPENDIX VIII: SUPPLEMENTAL INFORMATION ON LIMITATIONS AND UNCERTAINTIES OF EPA'S PERCHLORATE EXPOSURE ANALYSIS

To support its preliminary regulatory determination, EPA developed perchlorate exposure estimates using a novel methodology that merged biomonitoring data from CDC's National Health and Nutrition Examination Survey (CDC biomonitoring data) with perchlorate occurrence data from EPA's testing program. Using this methodology, EPA developed exposure estimates of daily perchlorate exposure from drinking water for various age and gender groups,[1] focusing largely on the perchlorate exposure of pregnant females because, EPA stated, the National Academies and EPA considered pregnant women and their fetuses to be the most sensitive population. In addition to the three limitations discussed in this report, reviewers noted additional limitations related to (1) the way in which participants were placed into groups that represented their potential for exposure to perchlorate from drinking water and food and (2) the adjustment—called a creatinine adjustment— that EPA had to make to estimate daily exposure to perchlorate from a single urine sample.

Limitations and Uncertainties Related to Grouping Participants According to Their Potential Exposure to Perchlorate from Drinking Water and Food

After merging the data sets, EPA separated study participants into three groups, according to their potential for exposure to perchlorate through drinking water and food: participants whose exposure to perchlorate most likely comes from food and water, participants whose exposure most likely comes only from food, and participants whose exposure is unknown because occurrence data were not available. Study participants residing in the same counties as public water systems that had at least one detection of perchlorate during the sample period were considered to be most likely exposed to perchlorate in both food and water. Study participants were placed in the group most likely exposed to perchlorate from food alone on the basis of meeting one or more of the following criteria: (1) they resided in counties where there were no quantified detections of perchlorate in public drinking water systems sampled under EPA's occurrence testing program, (2) they self-reported that they had not consumed public drinking water in the past 24 hours, or (3) they reported using a reverse osmosis filter at home (which can reduce perchlorate in drinking water). By placing participants into these two groups, EPA could estimate the relative contributions of food and water to perchlorate exposure. The remaining study participants were excluded from the analysis because EPA lacked data on perchlorate occurrence in drinking water to which they could be linked. However, as can be seen in the following examples, the data EPA had available to determine the group in which each study participant should be placed—exposure from food only or from food and water—and some criteria EPA applied in making these placements, introduce uncertainties into the exposure assessment that were not transparently acknowledged and explained in the agency's 2008 preliminary regulatory determination notice for perchlorate.

- EPA's testing program used a relatively insensitive minimum reporting level of 4 micrograms of perchlorate per liter of water.[2] As a result, EPA's analysis of perchlorate exposure of CDC biomonitoring study participants placed in the group assumed to be exposed to perchlorate from food only—characterized by the study's authors as the "most important category" for the purpose of investigating the dose of perchlorate in food—could overstate the number of participants who met the criterion to be placed into this group because the group could include people exposed to perchlorate at concentrations up to 4 micrograms of perchlorate per liter of water. Further, to the extent participants in the food-only group were exposed to perchlorate in drinking water at levels below the minimum reporting level, the estimates of exposure from food alone are overstated. According to one reviewer, the minimum reporting level of 4 micrograms per liter introduces an uncertainty into the analysis that needs to be appropriately characterized, indicating that its use would underestimate perchlorate exposure for participants in the food and water group and inappropriately place individuals in the food-only group who should have been placed in the food and water group.
- EPA's perchlorate exposure estimates for participants placed in the food-only and food and water groups may also be subject to error because data limitations did not enable EPA to link participants to the perchlorate occurrence test results for their public water systems; rather, the data enabled each participant to be matched only to perchlorate occurrence test results for a public water system in the county in which the participant resides.[3] As a result, some participants who were not actually exposed to perchlorate from their public water system may have been erroneously placed in the group exposed to perchlorate in food and water.
- As noted by some reviewers, EPA's criterion for placing participants in the group exposed to perchlorate from food alone if they reported they had not consumed tap water in the past 24 hours may have caused some participants to be erroneously placed in this group because these individuals may have nonetheless been exposed to perchlorate in drinking water that was used in the preparation of food, such as juice, soup, or coffee.
- Another source of uncertainty that reviewers raised is the extent to which the analysis of the merged data is nationally representative. If the results of the analysis are weighted toward areas with known high or low perchlorate concentration in drinking water rather than being nationally representative, this limitation and its effects should be identified. However, EPA's preliminary regulatory determination and the journal article on this analysis provide no information on the geographic coverage of the

merged data, such as the number of states and the number of public water systems[4] represented by the merged CDC biomonitoring and EPA occurrence data.

Limitations and Uncertainties Related to the Creatinine Adjustment

To estimate each participant's daily exposure to perchlorate from a single urine sample, EPA performed a creatinine adjustment on each sample. Making this adjustment required the following information for each participant: age, sex, weight, height, race, and lean body mass. As can be seen in the following examples, some of the assumptions that EPA applied in making these adjustments introduce uncertainties into the exposure assessment that were not transparently acknowledged and explained in the agency's 2008 preliminary regulatory determination notice for perchlorate.

- A key assumption EPA used in making the creatinine adjustment was that 100 percent of ingested perchlorate is eliminated in urine within a 24-hour period. However, this is not a settled issue in the scientific community. The 2010 journal article detailing the methodology EPA used identifies several studies that report lower percentages of perchlorate excreted in urine within 24 hours of exposure.[5] For example, the journal article states that "several recent" perchlorate exposure studies report that approximately 70 percent of a perchlorate dose is excreted in urine. The estimates in the studies cited range from 50 percent to 100 percent. The cited studies were published between 2000 and 2007 and, consequently, were available at the time EPA was developing its relative source contribution factor for its preliminary regulatory determination on perchlorate.
- Other limitations related to the creatinine adjustment that introduce uncertainty into the exposure estimates stem from differences among individuals in the timing of urine sample collections relative to when they most recently consumed food and water, as well as the expected variability among individuals' intake and excretion related to their dietary and water consumption patterns (e.g., some individuals may excrete more or less of what they eat or drink).

APPENDIX IX: CALCULATIONS FOR THE PERCHLORATE HEALTH REFERENCE LEVEL IN EPA'S 2008 PRELIMINARY REGULATORY DETERMINATION AND THE RELATED 2010 JOURNAL ARTICLE ON EPA'S EXPOSURE ANALYSIS METHODOLOGY

The exposure estimates for pregnant women that EPA presented in its 2008 preliminary determination for perchlorate differed from those presented in the 2010 journal article that presented the exposure analysis methodology EPA used to support its preliminary determination.[1] This difference occurred because, in response to reviewers' comments, the authors of the 2010 journal article removed four data points that were determined to be outliers—two of which were data points that corresponded to women. As a result, in the article they reported pregnant women's exposure to perchlorate from food alone at 0.198

micrograms per kilogram per day— compared with EPA's 0.263 estimate. This downward adjustment in the exposure estimate for pregnant women in 2010 resulted in a change in the relative source contribution estimate for perchlorate from 62 percent to 72 percent.[2] Nonetheless, the 2008 and 2010 exposure analyses both support the same health reference level of 15 parts per billion—the level EPA and other federal agencies agreed to. As shown in table 12, this consistency in the health reference level was maintained because key assumptions were changed in calculating the health reference level.

Table 12. Comparison of the Health Reference Level Calculations Used in EPA's Preliminary Perchlorate Regulatory Determination and Its Subsequent Article on the Perchlorate Exposure Methodology the Agency Used in Its Preliminary Regulatory Determination, Based on a Reference Dose of 0.7 Micrograms per Kilogram per Day

Health reference level (parts per billion, rounded)	=	[(Reference dose (micrograms per kilogram per day)	x	Body weight) (kilograms)	÷	Drinking water intake] (liters per day)	x	Relative source contribution[a] (expressed as a percentage)	Source and date
15	=	[(0.7	x	70)	÷	2]	x	62	EPA's preliminary regulatory determination, 2008
15	=	[(0.7	x	66)	÷	2.21]	x	72	*Journal of Exposure Science and Environmental Epidemiology* article, 2010

Source: GAO.

Notes: If EPA's 2008 calculation of the health reference level had retained the 62 percent relative source contribution factor but used the body weight and drinking water intake assumptions reflected in the journal article, the health reference level would have been 13 parts per billion. Conversely, using the 72 percent relative source contribution factor from the results reported in 2010 and the default estimates for body weight and drinking water intake that EPA used in 2008, the health reference level would be 18 parts per billion.

[a]EPA calculated the relative source contribution using the subtraction method.

As shown in the table, in its preliminary regulatory determination, EPA derived a relative source contribution of 62 percent on the basis of pregnant women's exposure to perchlorate from drinking water and used the agency's default assumptions for weight (70 kilograms) and drinking water intake (2 liters per day) to calculate the health reference level of 15 parts per billion. In the 2010 journal article, the authors support a health reference level of 15 parts per billion, but in this case, it is derived on the basis of a 72 percent relative source contribution that reflects both the data corrections the authors made in the exposure analysis and alternate assumptions for weight (66 kilograms) and drinking water intake (2.21 liters per day.) These calculations illustrate how sensitive the outcomes of risk assessment methodologies can be to sometimes minor changes in basic assumptions. The sensitivity of the outcome—which then drives policy decisions—to these changes in assumptions, underscores the need for transparency and consistency in EPA's selection of assumptions for a given risk assessment.

End Notes

[1] Pub. L. No. 93-523 (1974), codified as amended at 42 U.S.C. §§ 300f–300j-26 (2010).

[2] See, for example, *Oversight of Recent EPA Decisions Hearing Before the Senate Committee on Environment and Public Works,* 110th Cong. (Feb. 6, 2007) (statement of Sen. Barbara Boxer); *Endocrine Disrupting Chemicals in Drinking Water: Risks to Human Health and the Environment Hearing Before the House Committee on Energy and Commerce, Energy and Environment Subcommittee*, 111th Cong. (Feb. 25, 2010) (statement of Rep. Edward Markey); *Oversight Hearing on Public Health and Drinking Water Issues Hearing Before the Senate Committee on Environment and Public Works*, 112th Cong. (Feb. 2, 2011) (statements of Sen. Barbara Boxer, Sen. Benjamin Cardin, and Sen. Frank Lautenberg).

[3] Under the Safe Drinking Water Act, a drinking water contaminant is defined as any physical, chemical, biological, or radiological substance or matter in water.

[4] Hypothyroidism is a condition in which the body lacks sufficient thyroid hormone.

[5] GAO, *Perchlorate: Occurrence Is Widespread but at Varying Levels; Federal Agencies Have Taken Some Actions to Respond to and Lessen Releases*, GAO-10-769 (Washington, D.C.: Aug. 12, 2010).

[6] Unless otherwise stated, in this report we refer to the Safe Drinking Water Act as amended.

[7] In this report, we refer to *Federal Register* notices regarding EPA's regulatory determinations (notices) and EPA's regulatory determination support documents individually and collectively, as appropriate. When referring to these documents collectively, we use the term "regulatory determination documents."

[8] The public drinking water systems regulated by EPA and delegated states and tribes provide drinking water to 90 percent of U.S. residents. These public drinking water systems, which may be publicly or privately owned, serve at least 25 people or 15 service connections for at least 60 days per year. Private, individual household wells are not regulated by EPA. States can seek lead enforcement responsibility (called primacy) for public water systems if they adopt drinking water regulations that are no less stringent than the national primary drinking water regulations and meet other statutory and regulatory requirements.

[9] Under the statute, the first regulatory determination cycle was to have been completed— including notice and public comment—by August 2001; the second by August 2006; and the third by August 2011—requiring determinations on at least 15 contaminants. EPA did not meet the time frames in the statute but has made determinations on 21 contaminants as of February 2011.

[10] Since the 1996 amendments, EPA has finalized national primary drinking water regulations that it had previously proposed; the regulations address several previously unregulated contaminants.

[11] The planned regulation of a group of carcinogenic VOCs includes contaminants that are not regulated as well as contaminants that are regulated individually, such as benzene, TCE, PCE, and vinyl chloride. EPA stated it was moving toward establishing a drinking water standard to address a group of up to 16 toxic chemicals but has not provided details on how the planned regulation of a group of both regulated and currently unregulated contaminants would be handled. The EPA Administrator stated, for example, that the agency would be "working towards developing an update to the Safe Drinking Water Act [regulations]" to address this planned action.

[12] EPA could also make a regulatory determination for a contaminant not on the contaminant candidate list but has not done so.

[13] Under the statute, the candidate list was to be published in 1998 and every 5 years thereafter—therefore, in 2003 and 2008. EPA did not meet the time frame for publishing the second and third candidate lists.

[14] As discussed earlier, EPA reversed its 2008 preliminary determination to not regulate perchlorate in February 2011. The determination was made as an out-of-cycle action.

[15] Executive Order 12866 directs agencies, among other things, to "identify for the public, in a complete, clear, and simple manner, the substantive changes between the drafts submitted to [OMB] for review and the action subsequently announced," "identify for the public those changes in the regulatory action that were made at the suggestion or recommendation of [OMB]," and make available to the public the draft action as sent to OMB for review.

[16] EPA, "Drinking Water: Regulatory Determinations Regarding Contaminants on the Second Drinking Water Contaminant Candidate List—Preliminary Determinations," *Federal Register* (Washington, D.C.: May 1, 2007).

[17] Under its unregulated contaminants testing program, during each testing cycle, EPA has generally required public water systems serving more than 10,000 people, and a representative sample of those serving 10,000 or fewer people, to test for the presence of selected contaminants either two or four times during a consecutive 12-month period; testing frequency depends on whether the public water systems are served by ground or surface water. Under the statute, EPA was to promulgate the first list for testing up to 30 specified unregulated contaminants in 1999 and every 5 years thereafter. EPA published the first rule in 1999 but did not meet the time frame for the second rule, which was published in 2007. EPA proposed the third rule for the testing program in March 2011 and plans to issue a final rule in 2012.

[18] The 1996 amendments modified the authority EPA had under the 1986 amendments to require public water systems to test for unregulated contaminants, limiting the number of (1) unregulated contaminants for which some systems would be required to test and (2) public water systems serving populations of less than 10,000 that EPA could require to conduct such testing. (See app. II for more information on the testing programs established by the 1986 and 1996 amendments.)

[19] The IRIS database contains EPA's scientific position on the potential human health effects of more than 540 chemicals. According to EPA documents, health risk assessment data that would be deemed sufficient to characterize the potential health effects include assessments from the IRIS program, the Office of Pesticide Program in a Reregistration Eligibility Decision, the National Academies, or ATSDR.

[20] EPA defines a reference dose as an estimate (with uncertainty spanning perhaps an order of magnitude) of a daily oral exposure to the human population (including sensitive subpopulations) that is likely to be without an appreciable risk of deleterious effects during a lifetime. It can be derived from a no-observed-adverse-effect level, a lowest-observedadverse-effect level, or a benchmark dose, with uncertainty factors generally applied to reflect limitations of the data used.

[21] IRIS information includes the reference dose for noncancer health effects resulting from oral exposure, the reference concentration for noncancer health effects resulting from inhalation exposure, and the cancer assessment for both oral and inhalation exposure. Combined with specific situational exposure assessment information, the summary health hazard information in IRIS may be used as a source in evaluating potential public health risks from chemical substances found in the environment.

[22] For the carcinogens EPA has evaluated for regulatory determination, EPA's assessments have assumed that no level of exposure is safe, with the health reference level reflecting the concentration of the contaminant in drinking water equivalent to a one-in-a-million increased risk of getting cancer from a lifetime of exposure. EPA's cancer risk policy provides guidance on the data needed for EPA to assume that there is a threshold below which a carcinogenic contaminant does not pose a risk. For health problems other than cancer, EPA has generally posited that there is some safe level of exposure to a contaminant before adverse health effects occur.

[23] EPA, Office of Science and Technology and Office of Water, *Methodology for Deriving Ambient Water Quality Criteria for the Protection of Human Health* (Washington, D.C.: October 2000). This guidance document includes discussions of the two approaches the drinking water program has used for estimating the relative source contribution and criteria for selecting the appropriate approach. While the guidance was developed in the context of Clean Water Act water quality criteria, the Office of Water uses it in making its drinking water regulatory determinations as well as in establishing primary drinking water regulations.

[24] According to the guidance, adequate data are available data that describe central tendencies and high ends for relevant exposure sources and pathways. In addition, adequacy depends on whether the data are relevant to and representative of the population at risk; therefore, data may be adequate for some decisions and inadequate for others.

[25] EPA health advisories for regulated and unregulated contaminants are provided in EPA's *2009 Edition of the Drinking Water Standards and Health Advisories*, EPA 822-R-09-011 (Washington, D.C., October 2009).

[26] Although EPA announced in its July 2008 *Federal Register* notice that it would issue an updated health advisory for a 10th contaminant that the agency decided to not regulate— 1,3-dichloropropene—as of February 2011, the agency had not issued an updated health advisory for this contaminant.

[27] EPA's Science Advisory Board is a federal advisory committee established by Congress in 1978 with a broad mandate to advise the agency on technical matters. The Board has established several standing committees, including the Drinking Water Committee.

[28] *SAB Advisory on EPA's Draft Third Drinking Water Contaminant Candidate List (CCL3)*, EPA-SAB-09-011 (Washington, D.C., Jan. 29, 2009).

[29] National Drinking Water Advisory Council, *Report of the National Drinking Water Advisory Council Working Group on CCL & Six-Year Review, Recommendations on the CCL Regulatory Determination Protocol* (May 23, 2000). The council approved the report's recommendation that EPA use the protocol for making regulatory determinations on June 14, 2000.

[30] EPA uses occurrence data to identify human exposure to contaminants in drinking water.

[31] Perchlorate was one of the 40 contaminants that remained unaddressed because, as EPA explained, "additional information may be needed to more fully characterize perchlorate exposure" to support a regulatory determination during this period. In October 2008, EPA issued a preliminary determination on perchlorate, and in February 2011, EPA issued a final regulatory determination.

[32] The third candidate list includes 104 chemical and 12 microbial contaminants.

[33] EPA made substantial changes in its methodology for developing the third candidate list, including implementing a criteria-based screening process. The 40 contaminants from the previous candidate lists for which the agency had not made regulatory determinations were included among the contaminants that were subject to the screening process. Using the new screening process to re-evaluate these contaminants, EPA selected 18 to carry forward to the third candidate list and did not select the other 22.

[34] EPA had planned to test for RDX in the first testing cycle, but the agency did not require testing for this contaminant during the first cycle as it awaited refinement of the testing (analytic) method. EPA subsequently selected RDX again for testing under the second cycle, allocating 2 of the 60 available testing slots for the first two testing cycles to RDX.

[35] GAO, *Chemical Assessments: Low Productivity and New Interagency Review Process Limit the Usefulness and Credibility of EPA's Integrated Risk Information System*, GAO-08-440 (Washington, D.C.: Mar. 7, 2008).

[36] To make progress on its backlog of 70-plus ongoing assessments, the IRIS program identified a subset of assessments for completion, taking into account such factors as the assessment stage (with priority generally given to assessments in the later stages of development) and input from program offices.

[37] According to OMB documents, EPA is the only federal agency that conducts quantitative cancer assessments of chemicals. Moreover, the World Health Organization, ATSDR, and the California Environmental Protection Agency—entities EPA has cited as potential sources for health assessment data for regulatory determinations—do not have health assessments on RDX.

[38] The IRIS program provides an assessment time frame for standard assessments—23 months—but not for more complex or controversial assessments. Rather, the IRIS program indicates such assessments may take longer.

[39] As of February 2011, EPA had not finalized IRIS assessments for the six key contaminants that we reported in 2008 had been in progress for a number of years.

[40] For example, along with contaminants such as perchlorate and hexavalent chromium, RDX is on DOD's emerging contaminants program's action list. Under this program, a DOD team on "materials of evolving regulatory interest" focuses on a number of contaminants to develop actions to respond to such potential factors as health impacts, cleanup costs, compliance costs, readiness impacts, and facilities' life-cycle costs.

[41] Because terbufos degrades quickly in water, EPA is currently testing for its longer-lived degradate, terbufos sulfone.

[42] Public water systems served by surface water are to test four times, and systems served by groundwater are to test two times, during a consecutive 12-month period.

[43] Under the time frames set out in the statute for identification of contaminants to be tested, EPA would have issued the list of contaminants for the third testing cycle in 2009, and based on past time frames between publishing a list and beginning testing, public water utilities in the testing program sample would have started to test for up to 30 contaminants in January 2011. EPA has recently proposed the list and time frames for the third testing cycle (2012 through 2016), with testing to be conducted from January 2013 through December 2015.

[44] See footnotes 17 and 43.

[45] According to EPA officials, the agency expects to have complete data from the second testing cycle by the fall of 2011.

[46] EPA, "Drinking Water Contaminant Candidate List 3—Final Notice," 74 *Fed. Reg.* 51850 (Washington, D.C., Oct. 8, 2009).

[47] EPA is authorized to make regulatory determinations for contaminants that are not on the candidate list, but has not done so to date. For these 9 contaminants, the agency could make determinations for any of them either as part of a regulatory determination cycle or out of cycle, or could add any of them to the next candidate list.

[48] According to EPA testing program documentation, EPA believes assessment monitoring represents "the most effective and accurate survey approach," providing "a confidence level of 99 percent with an allowable error of plus or minus 1 percent." The documentation notes that using an approach with a greater margin of error, and the resulting smaller sample size, could cause the occurrence of the many contaminants that occur in 1 percent or less of drinking water systems on a national basis to be "missed entirely." The documentation further notes that even a small percentage of systems with detections can affect a significant population.

[49] The first cycle testing included approximately 3,800 systems and the second cycle about 4,200. Assessment monitoring includes all systems serving over 10,000 people and a representative sample of 800 smaller systems.

[50] The 29 contaminants tested under the screening survey include 14 contaminants from the first testing cycle and 15 from the second cycle.

[51] EPA, "Unregulated Contaminant Monitoring Regulation (UCMR) for Public Water Systems Revisions: Proposed Rule," *Federal Register* (Aug. 22, 2005).

[52] As noted earlier, we did not evaluate the testing program. As such, we did not examine issues such as the extent to which the tests used widely available technologies and national laboratory capacity to conduct more sophisticated tests.

[53] The 28 contaminants with proposed assessment monitoring testing are chemicals. For the other 2 contaminants—microbials (biological substances), which present different testing challenges than chemicals—EPA proposed "pre-screen testing" at 800 targeted, undisinfected groundwater wells from systems that serve 1,000 or fewer customers.

[54] In addition, according to its February 11, 2011, final regulatory determination notice for perchlorate, the agency developed a range of health reference levels for 14 life stages, some of which are lower than the minimum reporting level the testing program used for perchlorate.

[55] Determinations for five of the contaminants with minimum reporting levels greater than health reference levels were based on state testing data collected under the testing program pursuant to the 1986 amendments to the Safe Drinking Water Act.

[56] Three of the contaminants were part of the first testing cycle. The fourth—1,3- dichloropropene—was not officially part of the first testing cycle, but EPA conducted analyses on samples collected from 796 of the small drinking water systems that provided samples for assessment monitoring in conjunction with the first testing cycle.

[57] Because the second cycle of testing will support the third cycle of regulatory determinations in 2013, it is not known at this time the extent to which minimum reporting levels for the contaminants could exceed the health reference levels developed as part of the regulatory determination process.

[58] USGS, *Quality of Source Water from Public-Supply Wells in the United States, 1993- 2007* (Reston, Va., May 2010). This report is part of the USGS National Water-Quality Assessment (NAWQA) Program, which was established in 1991 to develop long-term, nationally consistent information on the quality of the nation's streams and groundwater, and thereby support scientifically sound decisions for water-quality management, regulation, and policy decisions. The objectives of NAWQA are to assess the status and trends of national water-quality conditions and to understand the factors and processes that govern those conditions.

[59] USGS's health benchmarks are comparable in concept to EPA's health reference levels, although USGS does not use its benchmarks for regulatory decisions. USGS uses EPA's maximum contaminant levels for regulated contaminants and develops health-based screening levels for unregulated contaminants using EPA Office of Water methodologies for establishing drinking water guidelines.

[60] In its 2010 report on its national water-quality assessment program from 1993 to 2007, USGS reports that its results for three contaminants included in EPA's testing program differ, "perhaps because the reporting levels in this study were about 170-fold to 670-fold lower than in the UCMR2 program [cycle 2 of EPA's testing program]."

[61] The Office of Water uses occurrence data from public water systems to assess exposure.

[62] The Superfund program, established under the Comprehensive Environmental Response, Compensation, and Liability Act of 1980, is to protect human health and the environment from the effects of hazardous substances. Under this program, EPA has the authority to (1) clean up hazardous waste sites and then seek reimbursement from the nonfederal parties legally responsible for contaminating them or (2) compel such responsible parties to clean up these sites.

[63] GAO-08-440.

[64] GAO, *High-Risk Series: An Update*, GAO-09-271 (Washington, D.C.: January 2009); and *High Risk Series: An Update*, GAO-11-278 (Washington, D.C.: February 2011).

[65] We are currently evaluating this program and EPA's progress in completing assessments in a separate review.

[66] A total of 24 chemical contaminants lacked health effects information; in 2008, EPA listed 19 contaminants with health effects information gaps and had removed 5 of the contaminants the agency originally identified as lacking health effects information in 1998. Of the 5 contaminants removed from the list, EPA completed IRIS assessments for 2 (boron and perchlorate), used the existing assessment for 1,1,2,2-tetrachloroethane that the IRIS program was in the process of updating, and used the assessment of the parent compound dacthal (DCPA) for the two dacthal (DCPA) degradates.

[67] In the *Federal Register* notice announcing its preliminary regulatory determinations for cycle 2, EPA reported using IRIS health risk assessments, in some cases augmented by subsequent Office of Pesticides Program assessments, for the 11 contaminants the agency determined had sufficient health risk data. EPA outlined its process for evaluating studies published after the assessments used were completed and for making needed adjustments on the basis of new data and to update the assessments to follow the agency's updated guidelines for carcinogen risk assessments. Office of Water officials told us that sometimes the office develops it own health assessment or uses other available assessments.

[68] EPA has yet to use an assessment by these entities to support its completed regulatory determinations. In its 2009 notice announcing its third candidate list, EPA indicated that the agency would need to evaluate the existing assessments from these other entities to determine whether the data on which they are based meet EPA data quality guidelines and are compatible with EPA risk assessment policies.

[69] GAO-08-440.

[70] At the time of this statement, EPA was operating under the framework of the 1986 amendments, rather than the 1996 amendments which are the focus of this section.

[71] USGS, *Quality of Source Water from Public-Supply Wells in the United States, 1993- 2007* (Reston, Va., May 2010).

[72] As noted earlier, USGS's health benchmarks are comparable in concept to EPA's health reference levels, although USGS does not use its benchmarks for regulatory decisions.

[73] This specific USGS report was not available at the time EPA made regulatory determinations on these contaminants, but most of the occurrence data in the report were available to EPA at the time the agency was developing regulatory determinations on these contaminants.

[74] EPA has used these USGS data for other purposes, such as evaluating chemicals for inclusion on its candidate lists.

[75] EPA, *Policy on Evaluating Health Risks to Children* (Washington, D.C.: Oct. 20, 1995).

[76] EPA, *EPA's Action Development Process-Guide to Considering Children's Health When Developing EPA Actions: Implementing Executive Order 13045 and EPA's Policy on Evaluating Health Risks to Children* (Washington, D.C.: Oct. 2006). EPA published this guide to consider children's health when developing actions, such as policies, regulatory actions, or risk assessments. Under this guidance, for example, if health risks will be considered to inform a policy or decision, the children's health policy applies and EPA should develop a separate assessment of risks to infants and children, evaluate health risks of infants and children in risk characterizations, or state clearly why this was not done. Office of Water officials told us that they did not use this guidance in the 2008 regulatory determinations because the office developed the framework for the analyses that supported the 2008 regulatory determinations in 2003.

[77] EPA, *Child-Specific Exposure Factors Handbook*, EPA/600/R-06/096F (Washington, D.C.: Sept. 2008); *A Framework for Assessing Health Risks of Environmental Exposures to Children*, EPA/600/R-05/093F (Washington, D.C.: Sept. 2006); and *Guidance on Selecting Age Groups for Monitoring and Assessing Childhood Exposures to Environmental Contaminants*, EPA/630/P-03/003F (Washington, D.C.: Nov. 2005).

[78] EPA/630/P-03/003F.

[79] EPA has identified 10 life stages for children from birth up to 21 years of age.

[80] According to EPA's guidance documents on assessing childhood exposures to environmental contaminants, EPA historically used a standard approach assuming a lifetime of constant exposure for an adult (EPA/630/P-03/003F).

[81] The council approach outlined geographic distribution using the USGS codes for the nation's 21 water-resources regions and the more than 200 water-resources subregions. The approach defined national occurrence as detections in more than 4 of the 21 USGS water-resources regions, and regional occurrence as detection in 2 or more of the subregions but fewer than 4 of the regions.

[82] Another state has found high levels of these contaminants near military facilities; in this case, the state has regulated these contaminants.

[83] According to EPA's June 2008 regulatory support document for the second cycle of regulatory determinations, the concentration of 2,4-dinitrotoluene detected through EPA's unregulated contaminants testing program was 333 micrograms per liter—more than 6,000 times the health reference level.

[84] According to EPA documents, these contaminants are released into the environment primarily from facilities that manufacture and process dinitrotouluene. A mixture of dinitrotoluene is used in automobile airbags, as an intermediate in the production of urethane foams, and for smokeless powders in the munitions industry.

[85] The regulatory determination support document also states that individuals with severely impaired kidney function might also be sensitive to boron exposure.

[86] EPA's boron health advisory provides, for a child weighing 10 kilograms, a 10-day (acute) health advisory level of 3 milligrams of boron per liter of drinking water and a "longer-term" level of 2 milligrams per liter. The advisory also provides, for an adult, a longer-term health advisory of 5 milligrams per liter and a lifetime health advisory of 5 milligrams per liter.

[87] The purpose presented in the health advisories themselves is as follows: "Health advisories serve as technical guidance to assist federal, state, and local officials responsible for protecting public health when emergency spills or contamination situations occur."

[88] These contaminants are boron; dacthal (DCPA) mono-acid degradate; dacthal (DCPA) di-acid degradate; 1,3-dichloropropene; 2,4-dinitrotoluene; 2,6-dinitrotoluene; 1,1,2,2- tetrachloroethane; manganese; sodium; and sulfate. EPA also issued a health advisory in conjunction with its preliminary determination on perchlorate.

[89] For example, in May 2008, the President of the Association of State Drinking Water Administrators told a congressional committee that "most states do not have the resources or expertise to independently develop drinking water regulations and therefore look to EPA to conduct the necessary research and collect the data and information needed to make regulatory decisions."

[90] As discussed earlier, in some cases the data represented detections in excess of minimum reporting levels.

[91] The Emergency Planning and Community Right-to-Know Act of 1986 requires EPA and states to collect data annually on releases and transfers of certain toxic chemicals from industrial facilities and make the data available to the public in the Toxics Release Inventory.

[92] These contaminants are aldrin; boron; dieldrin; DDE; 2,4-dinitrotoluene; 2,6- dinitrotoluene; hexachlorobutadiene; manganese; naphthalene; sodium; and sulfate.

[93] The five states, listed in order of the size of populations served by public water systems with manganese levels in excess of the health reference level, are Alabama (2.4 percent); Oregon (3.2 percent); New Jersey (9.1 percent); Illinois (14.7 percent); and California (27.2 percent).

[94] In six of the nine cases, EPA had no testing data that would identify exposure at the health reference level, and in three cases, EPA had only limited data on exposure at the health reference level.

[95] The contaminants are aldrin, dieldrin, hexachlorobutadiene, and sulfate.

[96] Dieldrin was banned by EPA in 1974 for most uses, except for the control of termites, and banned for all uses in 1987 because of concerns about environmental damage, harm to human health, and its ability to persist in the environment for decades.

[97] According to USGS, because water samples were collected prior to any treatment or blending that potentially could alter contaminant concentrations, the sampled groundwater represents the quality of the source water and not necessarily the quality of finished water ingested by the people served by these public wells. Water utilities, however, are not required to treat water for unregulated contaminants.

[98] The contaminants are DDE; 1,3-dichloropropene; 2,4-dinitrotoluene; 2,6-dinitrotoluene; and 1,1,2,2-tetrachloroethane. In the cases of 1,1,2,2-tetrachloroethane and some of the data for 1,3-dichloropropene, some minimum reporting levels exceeded health reference levels. For these data sources, EPA evaluated occurrence at the health reference level, but stated that because some reporting limits exceeded the thresholds of interest, the occurrence analyses may result in an underestimate of systems affected.

[99] In addition, in 2003, EPA made a regulatory determination on hexachlorobutadiene while a new risk assessment was underway. This IRIS assessment is still at the initial stage (draft development).

[100] The National Academies consists of four private, nonprofit organizations that advise the federal government on scientific and technical matters: the National Academy of Sciences, National Academy of Engineering, Institute of Medicine, and National Research Council.

[101] National Academies, *Health Implications of Perchlorate Ingestion* (Washington, D.C., 2005).

[102] EPA uses its Action Development Process—an agency administrative process for developing rules, policy statements, risk assessments, guidance documents, models that may be used in future rulemakings, statutorily mandated reports to Congress, and strategies related to regulations—to govern the agency's development and

review of regulatory determinations. Regulatory determinations are not rules; however, if a positive determination is made to regulate a contaminant, a rulemaking would follow.

[103] For the 2003 regulatory determination process, the participating intra-agency offices were the Office of Prevention, Pesticides, and Toxic Substances; the Office of Research and Development; the Office of General Counsel; the Office of Policy, Economics, and Innovation; and Regions 1, 3, 5, 8, and 9. For the 2008 regulatory determination process, in addition to these offices, the Office of Enforcement and Compliance Assistance also participated, and regional office participation included Regions 1, 2, 6, and 8.

[104] Regulatory determination options for each contaminant presented to senior management could include to regulate, not to regulate, or not to regulate but to issue a health advisory.

[105] While consensus is not required, officials said that the work groups try to obtain consensus on the selected option among senior management officials. Officials could not recall a work group that did not achieve consensus.

[106] The draft final determinations underwent the same concurrence and OMB review as the preliminary determinations before being finalized and published in the *Federal Register*.

[107] According to Office of Water officials, this group was primarily composed of the EPA Deputy Administrator and the Assistant Administrators for the Office of Water; Office of Research and Development; Office of Solid Waste and Environmental Remediation; Office of Policy, Economics, and Innovation; and Office of General Counsel. Select program office staff were involved in some non-decision-making activities of the group.

[108] The Perchlorate Interagency Working Group includes officials from OMB and the Office of Science and Technology Policy, both part of the Executive Office of the President; DOD; NASA; the Department of Energy; the Department of Health and Human Services' Food and Drug Administration and ATSDR; the Department of Agriculture; and the Department of the Interior.

[109] The Council on Environmental Quality, which is part of the Executive Office of the President, coordinates federal environmental efforts in the development of environmental policies and initiatives.

[110] This working group stemmed from interagency work groups that were established as early as 1998.

[111] We recently reported that EPA has not fully used the Office of Children's Health Protection and other child-focused resources. GAO, *Environmental Health: High-level Strategy and Leadership Needed to Continue Progress toward Protecting Children from Environmental Threats*, GAO-10-205 (Washington, D.C.: Jan. 28, 2010).

[112] According to officials from the Office of Children's Health Protection, their office was excluded from the small group of high-level officials and from providing input into the perchlorate preliminary determination because it had not been part of the original work group for the second cycle of regulatory determinations. Office of Children's Health Protection officials said they were involved only informally and intermittently in the regulatory determination work groups largely as a result of its staff and resource limitations.

[113] EPA established the Office of Children's Health Protection to implement Executive Order 13045: *Protection of Children from Environmental Health Risks and Safety Risks* (signed in 1997) directing all federal agencies to assess health and safety risks to children, coordinate research priorities on children's health, and ensure that their standards take into account special risks to children.

[114] See 42 U.S.C. § 300g-1(b)(1)(C) (requiring consideration of such factors, among others, in selecting unregulated contaminants for consideration); § 300g-1(b)(3)(C)(i)(V) (requiring EPA to publish an analysis of such effects when proposing a national primary drinking water regulation).

[115] According to an EPA official, the agreement was documented in an unattributed two-page white paper and faxed to EPA from the Council on Environmental Quality in early August 2008; EPA made some editorial changes to the document but did not alter the substance of the agreement.

[116] This drinking water equivalent level was calculated using EPA's default assumptions for adult weight—70 kilograms (154 pounds) and daily drinking water intake—2 liters (60 fluid ounces).

[117] EPA noted in the *Federal Register* notice that the final health reference level in the final regulatory determination for perchlorate would serve as the basis for the agency's health advisory for perchlorate and that thereafter it might be appropriate to use the health advisory value as a "to be considered" value in developing potential cleanup levels for perchlorate pursuant to Superfund. A federal drinking water standard for perchlorate could also serve as a possible basis for environmental cleanup goals, potentially affecting some federal agencies' as well as perchlorate manufacturers' cleanup requirements at Superfund and other contaminated sites.

[118] National Academies, *Health Implications of Perchlorate Ingestion* (Washington, D.C., 2005).

[119] The National Academies' recommended reference dose for perchlorate was 0.7 micrograms per kilogram of body weight per day.

[120] Under the typical process, at that time, the IRIS program would circulate an IRIS Toxicological Review support document in addition to the IRIS Summary among the consensus reviewers, generally allowing 45 days for the

review. The program manager noted in the request for the perchlorate IRIS assessment consensus review, however, that EPA would not develop and post an IRIS Toxicological Review support document as part of the IRIS assessment.

[121] The IRIS program officials could not provide copies of the consensus review memorandums. However, we were able to obtain comments from 5 of the 18 individual consensus reviewers.

[122] The final IRIS assessment for perchlorate was posted on the Web site 11 days after the IRIS program requested the consensus reviews.

[123] M. A. Greer et al., "Health Effects Assessment for Environmental Perchlorate Contamination: The Dose Response for Inhibition of Thyroidal Radioiodine Uptake in Humans," *Environmental Health Perspectives*, vol. 110, no. 9 (2002). The published article notes that the study was supported by a National Institutes of Health grant and the Perchlorate Study Group—whose members included such perchlorate manufacturers and users as Aerojet, American Pacific Corporation, Kerr-McGee Chemical, and Lockheed Martin.

[124] Uncertainty factors, also called variability factors, are default factors used in deriving a reference dose from experimental data. The factors are intended to account for such things as variation in susceptibility among the members of the human population, extrapolating from data obtained in a study with less-than-lifetime exposure, and uncertainty associated with extrapolation when the database is incomplete.

[125] In calculating the reference dose, the National Academies applied the uncertainty factor to the no-observed-effect level reported for the group of seven healthy, nonpregnant adult subjects in the study that received the lowest dose of perchlorate—the level selected by the National Academies as the point of departure for determining the reference dose. The noobserved-effect level represents an exposure level at which there is no statistically or biologically significant difference in the frequency or severity of any effect between the exposed population and its appropriate control.

[126] G. Ginsberg and D. Rice, "The NAS Perchlorate Review: Questions Remain about the Perchlorate RfD," *Environmental Health Perspectives*, vol. 113, no. 9 (2005).

[127] Perchlorate is actively transported into breast milk, where relatively high levels have been reported in the United States and Chile. The public health officials noted that there has been a lack of useful studies examining lactation as a route of exposure, resulting in a critical data gap that has added uncertainty to perchlorate risk assessment. Since the National Academies study, however, new studies have provided information on perchlorate in breast milk that could help reduce this uncertainty, several of which EPA references in its August 2009 perchlorate supplemental request for comments. 74 *Fed. Reg.* 41883 (Aug. 19, 2009).

[128] The public health officials cited a 1998 study in rats that suggests greater perchlorate toxicity to the thyroid from 90-day exposure than from 14-day exposure.

[129] The variables are (1) the uncertainty factors; (2) the dose used as the starting point, or point of departure, for the health-risk computation (i.e. whether to use the no-observedeffect level; the lowest-observed-adverse-effect level; or the benchmark dose level); and (3) the relative source contribution. EPA defended its decision not to use any additional uncertainty factors, stating that it was relying on the findings of the National Academies' study.

[130] California's and Massachusetts's limits are set in state drinking water regulations; New Jersey's limit was a proposed drinking water standard that lapsed in March 2010 when the state's newly appointed Commissioner of the Department of Environmental Protection decided to delay adopting a standard until EPA made its regulatory determination.

[131]T he 2005 National Academies report on perchlorate contained varying characterizations of sensitive subpopulations, sometimes referring to pregnant women and their fetuses alone as *the most sensitive subpopulation* and other times including infants in this designation. In addition, the National Academies identified developing children as a sensitive population and people with compromised thyroid function and people who are iodide-deficient as potentially sensitive populations.

[132] In fact, the Office of Water had developed a draft perchlorate section to include in the second cycle regulatory determination preliminary notice indicating regulation of perchlorate in drinking water was warranted.

[133] Key entities that might be impacted by EPA regulation of perchlorate include perchlorate manufacturers, public drinking water systems, companies and federal agencies that use perchlorate, and others that could be responsible for clean up of perchlorate contamination.

[134] While the Superfund program does not establish cleanup standards, it requires, among other things, that long-term cleanups meet applicable requirements based on standards for contaminants set under state or federal laws or regulations and in consideration of other guidance. A federal drinking water standard for perchlorate could supply cleanup values at sites where existing or potential sources of drinking water are contaminated, unless a more stringent state standard applies. See 40 C.F.R. § 300.430(e)(2)(i) (2010). For discussion of state

standards, see GAO, *Perchlorate: Occurrence Is Widespread but at Varying Levels; Federal Agencies Have Taken Some Actions to Respond to and Lessen Releases*, GAO-10-769 (Washington, D.C.: Aug. 12, 2010).

135 Biomonitoring is a method to identify exposure to environmental chemicals by measuring chemicals in people's tissue or body fluids such as blood and urine. CDC's National Health and Nutrition Examination Survey conducts biomonitoring on a nationally representative sample of the U.S. population every 2 years, testing blood and urine to identify exposure to hundreds of chemicals, including perchlorate.

136 EPA also considered the Food and Drug Administration's 2006 Total Diet Study, which reported in 2008 that at least one sample in 74 percent of the 285 foods tested contained detectable levels of perchlorate. The Food and Drug Administration's Total Diet Study is a food-monitoring study that the agency has conducted periodically since 1961. In 2005 and 2006, the Food and Drug Administration tested a set of 285 representative foods for perchlorate, among other things, making it the most comprehensive information available on the occurrence of perchlorate in the diet at that time. The 285 food items tested represented the major components of the American diet, such as dairy, meat, fruits, and vegetables. Using the analytical results for the food samples collected, Food and Drug Administration researchers calculated the estimated average daily perchlorate intake from food for the total U.S. population and for 14 age and gender subgroups. For more information on the Food and Drug Administration's Total Diet Study, see http://www.fda.gov/Food/FoodSafety /FoodContaminantsAdulteration/ TotalDietStudy/default.htm.

137 D. R. Huber et al., "Estimating perchlorate exposure from food and tap water based on U.S. biomonitoring and occurrence data," *Journal of Exposure Science and Environmental Epidemiology* (2010). Other contributing authors cited in the article include EPA contractors with Danya International and The Cadmus Group.

138 In addition to the three independent scientists who reviewed the analysis prior to EPA's use of the results in its preliminary determination for perchlorate, the analysis was reviewed by OMB and other members of the Perchlorate Interagency Working Group as well as reviewers for the *Journal of Exposure Science and Environmental Epidemiology*.

139 CDC's biomonitoring study included data on children 6 years of age and older and adults.

140 EPA, Office of Science and Technology and Office of Water, *Methodology for Deriving Ambient Water Quality Criteria for the Protection of Human Health* (Washington, D.C., October 2000). This guidance document includes discussions of the two approaches the drinking water program has used for estimating the relative source contribution and criteria for selecting the appropriate approach. While the guidance was developed in the context of Clean Water Act water quality criteria, the Office of Water uses it in making its drinking water regulatory determinations.

141 These calculations are based on EPA's exposure data at the 90th percentile, the exposure level EPA used in calculating the relative source contribution for perchlorate in its preliminary regulatory determination. Because of the anomalies in the exposure data for pregnant women at the 90th percentile that EPA used as the basis for its relative source contribution, we cannot calculate what the relative source contribution would have been if EPA had chosen to use the percentage method in its preliminary regulatory determination.

142 Using the exposure analysis data to derive the relative source contribution under the subtraction method resulted in a relative source contribution of 80 percent.

143 In this scenario, the calculated relative source contribution was 15.2 percent. However, we rounded this number to 20 percent per EPA's guidance that generally specifies a 20 percent floor and an 80 percent ceiling for the relative source contribution.

144 EPA, Office of Science and Technology and Office of Water, *Methodology for Deriving Ambient Water Quality Criteria for the Protection of Human Health* (Washington, D.C., October 2000).

145 According to its December 2010 *Exposure and Relative Source Contribution Analysis for Fluoride* (820-R-10-015), the Office of Water is in the process of considering refinements to the 2000 guidance for determining the relative source contribution for ambient water and drinking water. However, this document states that EPA developed the relative source contribution for fluoride using the 2000 guidance because those modifications were not available for the fluoride exposure assessment.

146 PBPK models are complex and involve numerous underlying assumptions that are imbedded in mathematical representations of the processes associated with how a contaminant behaves within, and is eliminated from, the body.

147 As discussed earlier in the report, the reference dose is an estimate of the total daily oral exposure to a contaminant that is not likely to cause an "appreciable risk of deleterious effects during a lifetime."

148 As discussed above, EPA agreed with other federal agencies on this conclusion and the rationale to support its regulatory determination using a health reference level of 15 parts per billion of perchlorate in drinking water, focusing on pregnant women and their fetuses as the most sensitive population.

[149] E. A. Merrill et al., "PBPK Model for Radioactive Iodide and Perchlorate Kinetics and Perchlorate-Induced Inhibition of Iodide Uptake in Humans," *Toxicological Sciences*, vol. 83 (2005).

[150] R. A. Clewell et al., "Perchlorate and radioiodide kinetics across life stages in the human; using PBPK models to predict dosimetry and thyroid inhibition and sensitive populations based on developmental stage," *Journal of Toxicology and Environmental Health Part A*, vol. 70, no. 5 (2007).

[151] The no-observed-effect level was the starting point (point of departure) to which the National Academies applied an uncertainty factor of 10 to arrive at its recommended reference dose.

[152] In its February 2011 final regulatory determination on perchlorate, EPA stated that iodide uptake inhibition may not reflect the relationship of the precursor event to adverse outcomes in newborns and infants, who may not have the iodide stores sufficient to offset the effects of reduced iodide uptake.

[153] EPA, "Inhibition of the Sodium-Iodide Symporter by Perchlorate: An Evaluation of Life Stage Sensitivity Using Physiologically-Based Pharmacokinetic (PBPK) Modeling," EPA/699/R-08/106A (Washington, D.C., May 2009).

[154] The objectives of this executive order are to enhance planning and coordination with respect to both new and existing regulations; to reaffirm the primacy of federal agencies in the regulatory decision-making process; to restore the integrity and legitimacy of regulatory review and oversight; and to make the process more accessible and open to the public.

[155] According to an EPA official, "August framework" refers to the agreement that was faxed to EPA from the Council on Environmental Quality that included this conclusion as a key component of the rationale EPA and other federal agencies agreed to in August 2008.

[156] Lisa P. Jackson, Administrator, Memorandum "Transparency in EPA's Operations," (Apr. 23, 2009). (Committing "to uphold the values of transparency and openness in conducting EPA operations" and asking employees to help "ensure EPA operates in full compliance with [the] principle" of transparency.)

[157] Executive Order 12866 requires agencies to, among other things, "[i]dentify for the public, in a complete, clear, and simple manner, the substantive changes between the draft submitted to [OMB] for review and the action subsequently announced," "[i]dentify for the public those changes in the regulatory action that were made at the suggestion or recommendation of [OMB]," and make available to the public the draft action as sent to OMB for review. To the extent EPA's regulatory determinations are subject to this executive order, including individual agency comments in the public record could be better aligned with the executive order's objectives.

End Notes for Appendix I

[1] In this report, we refer to *Federal Register* notices regarding EPA's regulatory determinations (notices) and EPA's regulatory determination support documents individually and collectively, as appropriate. When referring to these documents collectively, we use the term "regulatory determination documents." The support documents varied under the two regulatory determination cycles, with several support documents for each contaminant considered under the first cycle and a consolidated support document for each contaminant under the second cycle.

End Notes for Appendix II

[1] EPA, *Technical and Economic Capacity of States and Public Water Systems to Implement Drinking Water Regulations: Report to Congress*, EPA 810/R-93-001 (Washington, D.C., 1993).

[2] The National Academies consist of four private, nonprofit organizations that advise the federal government on scientific and technical matters: the National Academy of Sciences, National Academy of Engineering, Institute of Medicine, and National Research Council.

[3] Originally, maximum contaminant level goals were known as "recommended maximum contaminant levels."

[4] In 1914, the Public Health Service—now part of the U.S. Department of Health and Human Services—set the first federal drinking water standards. These standards addressed only bacteriological contaminants capable of causing contagious disease and applied only to water systems providing drinking water to interstate carriers, such as ships and trains. The Public Health Service revised and expanded its standards in 1925, 1946, and 1962. The 1962 standards—regulating about 30 contaminants including arsenic, lead, and turbidity—were the

most comprehensive federal drinking water standards in existence before the Safe Drinking Water Act of 1974.

[5] Trihalomethanes, the other contaminant subject to interim regulation, was not on the list of 83 because EPA said insufficient time had elapsed since its interim regulation was issued to support revision.

[6] EPA 810/R-93-001

[7] By 1993, the list of 83 had expanded to 85 as a result of the splitting of dichlorobenzene and radium into ortho-dichlorobenzene and para-dichlorobenzene and radium 226 and radium 228, respectively. In addition, EPA had substituted seven of the compounds on the original list.

[8] EPA promulgated national primary drinking water regulations for 5 radionuclides in 2000 and for arsenic in 2001. EPA considered sulfate as part of regulatory determination Cycle 1 and determined not to regulate it in 2003.

[9] In 1992, EPA issued a final regulation for hexachlorobenzene, which was not on the statutorily mandated list of contaminants, explaining that the contaminant was being regulated "because it has been found in drinking water and may cause adverse health effects." See 57 *Fed. Reg.* 31776, 31783 (1992).

[10] The amendments provided the EPA Administrator with discretion to require public water systems to test more frequently.

[11] EPA can also make a regulatory determination for a contaminant that is not on the contaminant candidate list but to date has not done so.

[12] Since the 1996 amendments, EPA has finalized national primary drinking water regulations that it had previously proposed; these address several previously unregulated contaminants.

End Note for Appendix V

[1] According to EPA's guidance document, the subtraction method can only be used when other exposure standards (criteria), such as Clean Air Act or Clean Water Act standards, are not relevant. In addition, adequate data to characterize the likelihood of exposure to relevant sources is required. See EPA, Office of Science Technology and Office of Water, *Methodology for Deriving Ambient Water Quality Criteria for the Protection of Human Health*, EPA-822-B-00-004 (Washington, D.C., October 2000).

End Notes for Appendix VI

[1] Manganese ore is smelted to produce an alloy used to improve the stiffness, hardness, and strength of steel; a number of manganese compounds are used in such products and applications as unleaded gasoline, water and wastewater treatment, matches, dry-cell batteries, fireworks, fertilizer, fungicides, varnish, and livestock food supplements.

[2] According to EPA, about 30 percent of community water system customers are served by groundwater systems.

[3] EPA stated in its regulatory determination support document providing the nationally extrapolated occurrence estimates for manganese that these estimates were not presented in the regulatory determination notice because national extrapolations for some of the contaminants considered for regulation "can be problematic" and that the National Inorganic and Radionuclide Survey data for manganese only represented groundwater public water systems. Thus, instead of reporting the nationally extrapolated occurrence estimates, EPA presented in the notice the unweighted sampling data for groundwater public water systems, highlighting exposure to 39,000 people.

[4] In 1993, in discussing strategies the agency might use under the 1986 statutory framework in determining whether to regulate chemicals, EPA stated that if nationally representative data showed stable occurrence levels at around 1/10th the maximum contaminant level goal with no indication of occurrence at higher levels, regulation may not be warranted. EPA, *Technical and Economic Capacity of States and Public Water Systems to Implement Drinking Water Regulations, Report to Congress*, EPA 810-R-93-001 (Washington, D.C., 1993). EPA identified strategies for a new proposed approach to select contaminants for regulation as EPA had nearly completed regulating the contaminants mandated under the 1986 amendments and was moving toward regulating contaminants selected by EPA.

[5] The USGS National Water-Quality Assessment program was initiated in 1991 as a longterm study of 59 significant watersheds and aquifers representing approximately two-thirds of the overall water use in the

United States in a similar proportion to the population served by public water systems. In its regulatory determination documents, EPA reported the manganese results for 36 of the 59 watersheds and aquifers tested in the 1990s that had undergone USGS quality assurance checks.

[6] USGS detections in source water exceeding the human health benchmark indicate levels of the contaminant that are a potential human health concern; however, these detections do not necessarily indicate a contaminant would be found in finished drinking water at that level because of the possibility of blending with other source water or treatment prior to consumption.

[7] The five states, listed in order of the size of populations served by public water systems with manganese levels in excess of the health reference level, are Alabama, Oregon, New Jersey, Illinois, and California.

[8] Superfund sites on EPA's National Priorities List represent those the agency has identified as among the most seriously contaminated sites, posing relatively high risks to human health or the environment for releases of hazardous substances. ATSDR states that although the total number of sites evaluated for manganese is not known, the possibility exists that the number of sites at which manganese is found may increase in the future as more sites are evaluated. ATSDR is the lead agency for conducting health assessments for Superfund sites and Resource Conservation and Recovery Act facilities.

[9] Manganese is on DOD's emerging contaminants watch list, along with other contaminants including dioxin, lead, cadmium, and cobalt.

[10] According to the National Academies, the lack of the development of the blood-brain barrier in infants and children results in higher uptake of manganese by the brain. EPA's IRIS assessment explains that when manganese is transported through the blood stream directly to the brain, it bypasses the liver and the opportunity for first-pass hepatic clearance, increasing manganese toxicity.

[11] In developing its oral reference dose for manganese, the Office of Water used the modifying factor of three that the IRIS assessment recommended for assessing exposure from drinking water to address these concerns associated with infants as well as to address the following risk issues: (1) individuals drinking water on an empty stomach, such as early in the morning, absorb more manganese; and (2) adverse health effects associated with a lifetime consumption of drinking water containing about 2 milligrams of manganese per liter were reported in a 1989 study. Use of the modifying factor had the effect of changing the reference dose from 0.14 milligrams per kilogram per body weight to 0.047 milligrams per kilogram per body weight.

[12] M. T. Brown and B. Foos, "Assessing Children's Exposures and Risks to Drinking Water Contaminants: A Manganese Case Study," *Human and Ecological Risk Assessment: An International Journal*, vol. 15, no. 5 (2009).

[13] Similarly, ATSDR's toxicological profile on manganese states that many reports indicate that oral exposure to manganese, especially from contaminated water sources, can produce significant health effects that have been most prominently observed in children and are similar to those observed from inhalation exposure. (Regarding inhalation exposure, ATSDR states that inhaled manganese is often transported directly to the brain before it is metabolized by the liver and can lead to such adverse neurological effects as mental disorders and manganism, whose symptoms include tremors, difficulty walking, and facial muscle spasms.) According to ATSDR, while an actual threshold at which manganese exposure produces neurological effects in humans has not been established, children consuming the same concentration of manganese in water as adults are ultimately exposed to a higher milligram-per-kilogram body weight ratio of manganese because of their lower body weight, higher consumption volume, and greater retention of manganese; children are also potentially more sensitive to manganese toxicity than adults. ATSDR concluded that, collectively, studies suggest that ingestion of water or food contaminated with manganese may result in adverse neurological effects.

[14] While the IRIS program initiated an update assessment for manganese in 2008, it has been temporarily suspended as the program focuses on assessments it has considered to be of high priority. As of May 2011, manganese is not reported in IRIS Track as an ongoing assessment.

[15] A 2010 study reported that low-level, chronic exposure to manganese from drinking water is associated with significant intellectual impairments in children. Its authors support revisiting the national and international guidelines for safe manganese in water, such as EPA's health advisory of 0.3 milligrams per liter. M. F. Bouchard et al., "Intellectual Impairment in School-Age Children Exposed to Manganese from Drinking Water," *Environmental Health Perspectives* (2010).

[16] The Office of Water applied the modifying factor the 1996 IRIS assessment recommended for assessing exposure from drinking water. See footnote 11.

[17] EPA has developed health assessment levels specifically for children routinely for many years for drinking water health advisories and has also, on occasion, developed them in assessing risks to children for primary drinking water regulations for other contaminants.

[18] Phytate, also called phytic acid, is the principal storage form of phosphorus in many plants, especially in wheat, rice, rye, barley, and beans. Phosphorus in this form is generally not bioavailable to humans because humans lack the digestive enzyme, phytase, required to separate phosphorus from the phytate molecule.

[19] National Academies, *Spacecraft Water Exposure Guidelines for Selected Contaminants* (Washington, D.C., 2006).

[20] In addition to setting health-based primary drinking water standards, under the Safe Drinking Water Act, EPA establishes secondary drinking water standards to address aesthetic problems, such as taste or odor. These standards are non-enforceable guidelines.

[21] This term refers to aesthetic problems, such as taste, color, or odor.

[22] In its regulatory determination notice and support document, EPA stated that the primary adverse effects seen in animals after chronic exposure to low doses of boron generally involve the testes and developing fetus. EPA also states that reproductive effects in males were noted in the chronic and subchronic animal studies discussed in the notice.

[2] As mentioned earlier, EPA data show that about 30 percent of public water system customers are served by groundwater systems.

[24] The American Water Works Research Foundation is now known as the Water Research Foundation.

[25] In this study, the foundation reported that with few exceptions, boron levels in untreated surface water represent a reasonable estimate of boron levels in treated water provided to customers of public water systems.

[26] ATSDR also stated that although the total number of National Priorities List sites evaluated for this substance is not known, the possibility exists that the number of sites at which boron is found may increase in the future as more sites are evaluated.

[27] The regulatory determination support document also states that individuals with severely impaired kidney function might also be sensitive to boron exposure.

[28] In its boron health advisory, for a child weighing 10 kilograms, EPA presented a 10-day (acute) health advisory level of 3 milligrams of boron per liter of drinking water; a "longer-term" health advisory level for children of 2 milligrams per liter; a "longer-term" health advisory level for an adult of 5 milligrams per liter; and a lifetime health advisory level for an adult of 5 milligrams per liter.

[29] The health reference level of 1.4 milligrams EPA developed for its regulatory determination is closer to drinking water standards or guidelines for boron that, according to EPA's Health Advisory, have been set by six states and the World Health Organization and that range from 0.6 milligrams to 1 milligram of boron per liter of drinking water than to the 5 milligrams per liter EPA developed for the related health advisory.

End Notes for Appendix VIII

[1] The study includes only children 6 years of age and older.

[2] In 2002, Massachusetts's tests for perchlorate were sensitive enough to detect concentrations of perchlorate of less than 1 part per billion. Tests by the Departments of Defense and Energy have also detected concentrations of perchlorate in drinking water and groundwater of less than 1 part per billion.

[3] The CDC biomonitoring data and EPA occurrence data could be merged only at the county level because more specific location data, such as home addresses—which might have been useful for more accurately assigning individuals residing in counties served by multiple water systems to the correct group—could not be used because of privacy concerns.

[4] Some study participants may obtain their drinking water from private wells. In the exposure analysis, these individuals would be linked with the perchlorate test for a public water system in their county.

[5] D. R. Huber et al., "Estimating perchlorate exposure from food and tap water based on U.S. biomonitoring and occurrence data," *Journal of Exposure Science and Environmental Epidemiology* (2010).

End Notes for Appendix IX

[1] D. R. Huber et al., "Estimating perchlorate exposure from food and tap water based on U.S. biomonitoring and occurrence data," *Journal of Exposure Science and Environmental Epidemiology* (2010).

[2] The 72 percent relative source contribution is calculated using the adjusted exposure number for pregnant women that was reported in the journal article.

In: Drinking Water and Unregulated Contaminants
Editor: Cynthia Romero
ISBN: 978-1-63321-164-3
© 2014 Nova Science Publishers, Inc.

Chapter 3

SAFE DRINKING WATER ACT (SDWA): A SUMMARY OF THE ACT AND ITS MAJOR REQUIREMENTS*

Mary Tiemann

SUMMARY

This report summarizes the Safe Drinking Water Act (SDWA) and its major programs and regulatory requirements. This report includes the drinking water security provisions added to the SDWA by the Public Health Security and Bioterrorism Preparedness and Response Act of 2002 (P.L. 107-188), and lead reduction provisions as amended by P.L. 111-380. It also outlines amendments made in December 2013 by P.L. 113-64 (H.R. 3588) to explicitly exempt fire hydrants from coverage under the act's lead plumbing restrictions.

The SDWA, Title XIV of the Public Health Service Act, is the key federal law for protecting public water supplies from harmful contaminants. First enacted in 1974 and substantially amended in 1986 and 1996, the act is administered through programs that establish standards and treatment requirements for public water supplies, control underground injection of wastes, finance infrastructure projects, and protect sources of drinking water. The 1974 law established the current federal-state arrangement in which states may be delegated primary implementation and enforcement authority for the drinking water program. The state-administered Public Water Supply Supervision (PWSS) Program remains the basic program for regulating the nation's public water systems, and 49 states have assumed this authority.

The last major reauthorization of the act was done through the Safe Drinking Water Act Amendments of 1996 (P.L. 104-182), which generally authorized appropriations for SDWA programs through FY2003. As with other EPA-administered statutes having expired funding authority, Congress has continued to appropriate funds for the ongoing SDWA programs.

* This is an edited, reformatted and augmented version of a Congressional Research Service publication, CRS Report for Congress RL31243, prepared for Members and Committees of Congress, from www.crs.gov, dated February 5, 2014.

In addition to reviewing key programs and requirements of the SDWA, this report includes statistics on the number and types of regulated public water systems. It also provides a table that lists all major amendments, with the year of enactment and public law number, and that cross-reference sections of the act with the major *U.S. Code* sections of the codified statute.

INTRODUCTION

The Safe Drinking Water Act (SDWA), Title XIV of the Public Health Service Act, is the key federal law for protecting public water supplies from harmful contaminants. First enacted in 1974 and substantially amended in 1986 and 1996, the act is administered through programs that establish standards and treatment requirements for public water supplies, control underground injection of wastes, finance infrastructure projects, and protect sources of drinking water. The 1974 law established the current federal-state arrangement in which states may be delegated primary implementation and enforcement authority for the drinking water program. The state-administered Public Water Supply Supervision (PWSS) Program remains the basic program for regulating the nation's public water systems, and 49 states have assumed this authority. The 1996 SDWA amendments generally authorized appropriations for SDWA programs through FY2003. *Table 1* identifies the original enactment and subsequent amendments.

Table 1. Safe Drinking Water Act and Amendments (codified generally as 42 U.S.C. 300f-300j)

Year	Act	Public Law Number
1974	Safe Drinking Water Act of 1974	P.L. 93-523
1977	Safe Drinking Water Act Amendments of 1977	P.L. 95-190
1979	Safe Drinking Water Act Amendments	P.L. 96-63
1980	Safe Drinking Water Act Amendments	P.L. 96-502
1986	Safe Drinking Water Act Amendments of 1986	P.L. 99-339
1988	Lead Contamination Control Act of 1988	P.L. 100-572
1996	Safe Drinking Water Act Amendments of 1996	P.L. 104-182
2002	Public Health Security and Bioterrorism Preparedness and Response Act of 2002	P.L. 107-188
2011	Reduction of Lead in Drinking Water Act	P.L. 111-380
2013	Community Fire Safety Act of 2013	P.L. 113-64

This report summarizes the act's major provisions, programs, and requirements. It also provides selected statistics on the universe of regulated public water systems, and lists references for further information on the act and its implementation. *Table 2*, located at the end of this report, cites the major *U.S. Code* sections of the act and the equivalent sections of the statute.

BACKGROUND

As indicated by *Table 1*, the Safe Drinking Water Act has been amended several times since enactment of the Safe Drinking Water Act of 1974 (P.L. 93-523). Congress passed this law after nationwide studies of community water systems revealed widespread water quality problems and health risks resulting from poor operating procedures, inadequate facilities, and uneven management of public water supplies in communities of all sizes. The 1974 law gave the EPA substantial discretionary authority to regulate drinking water contaminants and gave states the lead role in implementation and enforcement.

The first major amendments (P.L. 99-339), enacted in 1986, were largely intended to increase the pace at which the EPA regulated contaminants and to increase the protection of ground water. From 1974 until 1986, the EPA had regulated just one additional contaminant beyond the 22 standards previously developed by the Public Health Service. The 1986 amendments required the EPA to (1) issue regulations for 83 specified contaminants by June 1989 and for 25 more contaminants every three years thereafter, (2) promulgate requirements for disinfection and filtration of public water supplies, (3) ban the use of lead pipes and lead solder in new drinking water systems, (4) establish an elective wellhead protection program around public wells, (5) establish a demonstration grant program for state and local authorities having designated sole-source aquifers to develop ground water protection programs, and (6) issue rules for monitoring injection wells that inject wastes below a drinking water source. The amendments also increased EPA's enforcement authority.

Congress again amended SDWA with the Lead Contamination Control Act of 1988 (P.L. 100- 572). These provisions were intended to reduce exposure to lead in drinking water by requiring the recall of lead-lined water coolers, and requiring the EPA to issue a guidance document and testing protocol for states to help schools and day care centers identify and correct lead contamination in school drinking water.

After the regulatory schedule mandated in the 1986 amendments proved to be unworkable for the EPA, states, and public water systems, the 104th Congress made sweeping changes to the act with the SDWA Amendments of 1996 (P.L. 104-182). As over-arching themes, the amendments targeted resources to address the greatest health risks, added some regulatory flexibility, provided funding for federal drinking water mandates, and aimed to improve water system compliance capacity. Congress revoked the requirement that the EPA regulate 25 new contaminants every three years and created a risk-based approach for selecting contaminants for regulation.

Among other changes, Congress added some flexibility to the standard setting process, required the EPA to conduct health risk reduction and cost analyses for new rules, authorized a drinking water state revolving loan fund (DWSRF) program to help water systems finance projects needed to meet SDWA requirements, added programs to improve small system compliance, expanded consumer information requirements, increased the act's focus on pollution prevention through a voluntary source water protection program, and streamlined the act's enforcement provisions. P.L. 104-182 extended authorizations for appropriations under the act through FY2003.

In 2002, several drinking water security provisions were added to the SDWA through the Public Health Security and Bioterrorism Preparedness and Response Act of 2002 (P.L.

107-188). Title IV of the act included requirements for community water systems serving more than 3,300 individuals to conduct vulnerability assessments and prepare emergency response plans. The law also required the EPA to conduct research on preventing and responding to terrorist or other attacks.

REGULATED PUBLIC WATER SYSTEMS

Federal drinking water regulations apply to the approximately 152,700 privately and publicly owned water systems that provide piped water for human consumption to at least 15 service connections or that regularly serve at least 25 people. These water systems vary greatly in size and type, ranging from large municipal systems to homeowner associations, schools, and campgrounds.

More than 51,350 of the regulated public water systems are *community water systems* (CWS) that serve the same residences year-round. These water systems provide water to approximately 299 million people. All federal regulations apply to these systems. Most community water systems (82%) are relatively small, serving 3,300 people or fewer. Despite this large percentage, these systems provide water to just 9% of the total population served by community water systems. In contrast, 8% of CWS serve populations of 10,000 or more, but provide water to 82% of the population served (more than 246 million individuals). Among the community water systems, 77% rely on ground water and 23% rely on surface water.

Another 18,178 public water systems are *non-transient non-community water systems* (NTNCWS), such as schools or factories, which have their own water supply and generally serve the same individuals for more than six months but not year-round. Most drinking water regulations apply to these systems. Of these water systems, 99% are small or very small, and provide water to 83% of the population served by these systems.

Nearly 83,200 other public water systems are *transient non-community water systems* (TNCWS), such as campgrounds and gas stations, which provide their own water to transitory customers. Only regulations for contaminants that pose immediate health risks apply to these systems.[1]

Approximately 95,700 of the nearly 101,400 non-community water systems (transient and non-transient systems combined) serve 500 or fewer people. These statistics give some insight into the scope of technological, economic, and managerial challenges small public water systems may face in meeting federal drinking water regulations. (*Figure 1* provides statistics on community water systems, non-transient non-community water systems, and transient non-community water systems.)

NATIONAL DRINKING WATER REGULATIONS

A key component of SDWA is the requirement that the EPA promulgate national primary drinking water regulations for contaminants that may pose health risks and that are likely to be present in public water supplies. Section 1412 instructs the EPA on how to select contaminants for regulation and specifies how the EPA must establish regulations once a contaminant has been selected. The regulations apply to the roughly 168,000 privately and publicly owned water systems that provide piped water for human consumption to at least 15 service connections or that regularly serve at least 25 people. The EPA has issued regulations for more than 90 contaminants, including regulations setting new standards for drinking water disinfectants and their byproducts and for microbial contaminants, a regulation establishing a standard for uranium in drinking water, and a regulation revising the standard for arsenic.

		Water System Population Size Category					
		Very Small 500 or less	Small 501-3,300	Medium 3,301-10,000	Large 10,001-100,000	Very Large >100,000	Totals
CWS	# Systems	28,3462	13,737	4,936	3,802	419	51,356
	Pop. Served	4,763,672	19,661,787	28,737,564	108,770,014	137,283,104	299,216,141
	% of Systems	55%	27%	10%	7%	1%	100%
	% Of Pop.	2%	7%	10%	36%	46%	100%
NTNCWS	# Systems	15,461	2,566	132	18	1	18,178
	Pop. Served	2,164,594	2,674,694	705,320	441,827	203,000	6,189,435
	% of Systems	85%	14%	1%	0%	0%	100%
	% Of Pop.	35%	43%	11%	7%	3%	100%
TNCWS	# Systems	80,347	2,726	92	13	1	83,179
	Pop. Served	7,171,054	2,630,931	514,925	334,715	2,000,000	12,651,625
	% of Systems	97%	3%	0%	0%	0%	100%
	% Of Pop.	57%	21%	4%	3%	16%	100%
	Total # of Systems	124,270	19,029	5,160	3,833	421	152,713

Source: U.S. Environmental Protection Agency, Fiscal Year 2011 Drinking Water and Ground Water Statistics, EPA 816-R-13-003, March 2013, p. 8, http://water.epa.gov/scitech/datait/databases/drink/sdwisfed/upload/ epa816r13003.pdf.

Notes: EPA has established three broad categories of public water systems. A community water system (CWS) serves the same population year-round. A non-transient non-community water system (NTNCWS) regularly supplies water to at least 25 of the same people at least six months per year, but not year-round (e.g., schools, factories, office buildings, and hospitals that have their own water systems. Transient non-community water systems (TNCWS) provide water in places where people do not remain for long periods of time, such as gas stations and campgrounds.

Figure 1. Public Water System Statistics. (water systems regulated under the SDWA)

Contaminant Selection and Regulatory Schedules

Section 1412, as amended in 1996, directs the EPA to select contaminants for regulatory consideration based on occurrence, health effects, and meaningful opportunity for health risk reduction. Starting in 1998 and then every five years, the EPA must publish a list of contaminants that may warrant regulation. As of 2001, and every five years thereafter, the EPA must determine whether or not to regulate at least five of the listed contaminants. The act requires the EPA to evaluate contaminants that present the greatest health concern and to regulate those contaminants that occur at concentration levels and frequencies of public health concern. The amendments also included schedules for the EPA to complete regulations for specific contaminants (i.e., radon, arsenic, disinfectants and disinfection byproducts and *Cryptosporidium*).

Standard Setting

For each contaminant that the EPA determines requires regulation, the EPA must set a nonenforceable maximum contaminant level goal (MCLG) at a level at which no known or anticipated adverse health effects occur and which allows an adequate margin of safety. The EPA must then set an enforceable standard, a maximum contaminant level (MCL), as close to the MCLG as is "feasible" using best technology, treatment techniques, or other means available (taking costs into consideration). The agency generally sets standards based on technologies that are affordable for large communities; however, as amended by P.L. 104-182, the act requires EPA, when issuing a regulation for a contaminant, to list any technologies or other means that comply with the MCL and that are affordable for small public water systems serving populations of 10,000 or fewer. If the EPA does not identify "compliance" technologies that are affordable for small systems, then it must identify small system "variance" technologies or other means that may not achieve the MCL but are protective of public health. The 1996 amendments authorized the EPA to set a standard at other than the feasible level if the feasible level would lead to an increase in health risks by increasing the concentration of other contaminants or by interfering with the treatment processes used to comply with other SDWA regulations. In such cases, the standard or treatment techniques must minimize the overall health risk. Also, when proposing a regulation, the EPA now must publish a determination as to whether or not the benefits of the standard justify the costs. If the EPA determines that the benefits do not justify the costs, the agency may, with certain exceptions, promulgate a standard that maximizes health risk reduction benefits at a cost that is justified by the benefits.[2]

New regulations generally become effective three years after promulgation. Up to two additional years may be allowed if the EPA (or a state in the case of an individual system) determines the time is needed for capital improvements. (Section 1448 outlines procedures for judicial review of EPA actions involving the establishment of SDWA regulations and other final EPA actions.)

Risk Assessment

The 1996 amendments also added risk assessment and risk communication provisions to SDWA. When developing regulations, the EPA is required to (1) use the best available, peer-reviewed science and supporting studies and data; and (2) make publicly available a risk assessment document that discusses estimated risks, uncertainties, and studies used in the assessment. When proposing drinking water regulations, the EPA must publish a health risk reduction and cost analysis (HRRCA). The EPA may promulgate an interim standard without first preparing this benefit-cost analysis or making a determination as to whether the benefits of a regulation would justify the costs if the Administrator determines that a contaminant presents an urgent threat to public health.

Variances and Exemptions

In anticipation that some systems, particularly smaller ones, could have difficulty complying with every regulation, Congress included in SDWA provisions for variances and exemptions. Section 1415 authorizes a state to grant a public water system a *variance* from a standard if raw water quality prevents meeting the standard despite application of best technology, and the variance does not result in an unreasonable risk to health.

Subsection 1415(e) authorizes variances specifically for small systems, based on application of best affordable technology. When developing a regulation, if the EPA cannot identify a technology that meets the standard and is affordable for small systems, the EPA must identify variance technologies that are affordable but do not necessarily meet the standard. In cases where the EPA has identified variance technologies, then states may grant small system variances to systems serving 3,300 or fewer persons if the system cannot afford to comply with a standard (through treatment, an alternative water source, or restructuring) and the variance ensures adequate protection of public health. A state may then grant a variance to a small system, allowing the system to use a variance technology to comply with a regulation. With EPA approval, states also may grant variances to systems serving between 3,301 and 10,000 persons. Variances are not available for microbial contaminants. *The EPA has determined that affordable compliance technologies are available for all existing standards. Thus, small system variances are not available.*

Section 1416 authorizes states to grant public water systems temporary *exemption*s from standards or treatment techniques if a system cannot comply for other compelling reasons (including costs). An exemption is intended to give a water system more time to comply with a regulation and can be issued only if it will not result in an unreasonable health risk. A qualified system may receive an exemption for up to three years beyond the compliance deadline. Systems serving 3,300 or fewer persons may receive a maximum of three additional two-year extensions, for a total exemption duration of nine years.

State Primacy

Section 1413 authorizes states to assume primary oversight and enforcement responsibility (primacy) for public water systems. To assume primacy, states must adopt regulations at least as stringent as national requirements, develop adequate procedures for enforcement (including conducting monitoring and inspections), adopt authority for administrative penalties, and maintain records and make reports as EPA may require. States also must develop a plan for providing safe drinking water under emergency circumstances. Currently, 55 of 57 states and territories have primacy authority for the public water system supervision (PWSS) program.

Under Section 1443, Congress authorized appropriations of $100 million annually for the EPA to make grants to states to administer the PWSS program. This section directs the EPA, in accordance with regulations, to allot the sums among the states "on the basis of population, geographical area, number of public water systems, and other relevant factors." States are authorized to use a portion of their drinking water state revolving fund grant (under Section 1452) to cover the costs of administering the PWSS program.

Enforcement, Consumer Information, and Citizen Suits

The Safe Drinking Water Act requires public water systems to monitor their water supplies to ensure compliance with drinking water standards and to report monitoring results to the states. States review monitoring data submitted by public water systems, and also conduct their own monitoring, to determine system compliance with drinking water regulations. The EPA monitors public water system compliance primarily by reviewing the violation data submitted by the states.

Section 1414 requires that, whenever the EPA finds that a public water system in a state with primary enforcement authority does not comply with regulations, the agency must notify the state and the system and provide assistance to bring the system into compliance. If the state fails to commence enforcement action within 30 days after the notification, the EPA is authorized to issue an administrative order or commence a civil action. In a nonprimacy state, the EPA must notify an elected local official (if any has jurisdiction over the water system) before commencing an enforcement action against the system.

The 1996 amendments strengthened enforcement authorities, streamlined the process for issuing federal administrative orders, increased administrative penalty amounts, made more sections of the act clearly subject to EPA enforcement, and required states (as a condition of primacy) to have administrative penalty authority. The amendments also provided that no enforcement action may be taken against a public water system that has a plan to consolidate with another system.[3]

Consumer Information and Reports

Enforcement provisions also require public water systems to notify customers of violations of drinking water standards or other requirements, such as monitoring and reporting requirements. Systems must notify customers within 24 hours of any violations that have the potential to cause serious health effects. Additionally, community water systems must mail to all customers an annual "consumer confidence report" on contaminants detected in their drinking water. States are required to prepare annual reports on the compliance of public water systems and to make summaries available to the EPA and the public; the EPA must prepare annual national compliance reports (§1414(c)).

Citizen Suits

Section 1449 provides for citizens' civil actions. Citizen suits may be brought against any person or agency allegedly in violation of provisions of the act, or against the EPA Administrator for alleged failure to perform any action or duty which is not discretionary.

COMPLIANCE IMPROVEMENT PROGRAMS

The 1996 amendments added two state-administered programs aimed at improving public water system compliance with drinking water regulations: the operator certification program and the capacity development program. Section 1419 required states to adopt programs for training and certifying operators of community and non-transient non-community systems (e.g., schools and workplaces that have their own wells). The EPA is required to withhold 20% of a state's annual DWSRF grant, unless the state has adopted and implements an operator certification program. Relatedly, Section 1420 required states to establish capacity development programs, also based on EPA guidance. Congress specified that the programs must include (1) legal authority to ensure that new systems have the technical, financial, and managerial capacity to meet SDWA requirements; and (2) a strategy to assist existing systems that are experiencing difficulties to come into compliance. The EPA is required to withhold a portion of SRF grants from states that do not have capacity development strategies. The agency has not had to withhold funds under either of these programs.

GROUND WATER PROTECTION PROGRAMS

Most public water systems rely on ground water as a source of drinking water, and Part C of the act focuses on ground water protection.[4] Section 1421 authorized the establishment of state underground injection control (UIC) programs to protect underground sources of drinking water (USDWs). In 1977, the EPA issued mandated regulations that contained minimum requirements for state UIC programs to prevent underground injection that endangers drinking water sources, and that required states to prohibit any underground injection not authorized by state permit. The law specified that the regulations could not interfere with the underground injection of brine from oil and gas production or recovery of oil unless underground sources of drinking water would be affected.[5]

Section 1422 authorized affected states to submit plans to the EPA for implementing UIC programs and, if approved, to assume primary enforcement responsibility. If a state's plan has not been approved, or the state has chosen not to assume program responsibility, then the EPA must implement the program (§1423). For oil and gas injection operations only, states with UIC programs are delegated primary enforcement authority without meeting EPA regulations under Section 1421, provided states demonstrate that they have an effective program that prevents underground injection that endangers drinking water sources (§1425, added in 1980). EPA has delegated primacy for all classes of wells to 34 states; it shares implementation responsibility in six states, and implements the UIC program for all well classes in 10 states.

To implement this program, EPA placed UIC wells in five classes based on similarity in the fluids injected, and construction, injection depth, design, and operating techniques, and issued regulations that establish performance criteria for each class.[6] In 2008, the agency proposed a rule to establish a new Class VI well for use in the underground injection of carbon dioxide. Class VI wells would be used for the long-term geologic sequestration of carbon dioxide as a tool for mitigating greenhouse gas emissions from coal-fired power plants and other large stationary sources of carbon dioxide.

A separate ground water protection provision, Subsection 1424(e), authorizes the EPA to make determinations, on EPA's initiative or upon petition, that an aquifer is the sole or principal drinking water source for an area. In areas that overlie a designated sole-source aquifer, no federal funding may be committed for projects that the EPA determines may contaminate such an aquifer. Any person may petition for sole source aquifer designation. Nationwide, the EPA had designated 73 sole source aquifers.

The act contains three other state programs aimed specifically at protecting ground water. Added in 1986, Section 1427 established procedures for demonstration programs to develop, implement, and assess critical aquifer protection areas already designated by the Administrator as sole source aquifers. Section 1428, also added in 1986, established an elective state program for protecting wellhead areas around public water system wells. If a state established a wellhead protection program by 1989, and the EPA approved the state's program, then the EPA may award grants covering between 50% and 90% of the costs of implementing the program. In 1996, Congress added Section 1429, authorizing the EPA to make 50% grants to states to develop programs to ensure coordinated and comprehensive protection of ground water within the states. For these programs, appropriations were authorized through FY2003 as follows: $15 million per year for Section 1427, $30 million per year for Section 1428, and $15 million per year for Section 1429.

Source Water Assessment and Protection Programs

The 1996 amendments expanded the act's pollution prevention focus to embrace protection of surface water, as well as ground water. Section 1453 required the EPA to publish guidance for states to implement source water assessment programs that delineate boundaries of the areas from which systems receive water, and identify the origins of contaminants in those areas to determine systems' susceptibility to contamination. States with approved assessment programs may adopt alternative monitoring requirements for water systems, as provided for in Section 1418.

Section 1454 authorized a source water petition program based on voluntary partnerships between state and local governments. States may establish a program under which a community water system or local government may submit a petition to the state requesting assistance in developing a voluntary source water quality protection partnership to (1) reduce the presence of contaminants in drinking water, (2) receive financial or technical assistance, and (3) develop a long-term source water protection strategy. This section authorized $5 million each year for grants to states to support petition programs. Also, states may use up to 10% of their DWSRF grant to support various source water protection activities including the petition program.

State Revolving Funds

In 1996, Congress authorized a drinking water state revolving loan fund (DWSRF) program to help systems finance improvements needed to comply with SDWA regulations (§1452). The EPA is authorized to make grants to states to capitalize DWSRFs, which states then may use to make loans to public water systems. States must match 20% of the federal grant. Grants are allotted based on the results of needs surveys.[7] Each state and the District of Columbia must receive at least 1% of the appropriated funds.

Drinking water SRFs may be used to provide loans for expenditures that the EPA has determined will facilitate compliance or significantly further the act's health protection objectives. States must make available 15% of their annual allotment for loan assistance to systems that serve 10,000 or fewer persons, to the extent that funds can be obligated for eligible projects. States may use up to 30% of their DWSRF grant to provide loan subsidies (including forgiveness of principal) to help economically disadvantaged communities. Also, states may use a portion of funds for technical assistance, source water protection and capacity development programs, and for operator certification.

The law authorized appropriations of $599 million for FY1994 and $1 billion per year for FY1995 through FY2003 for DWSRF capitalization grants. The EPA was directed to reserve, from annual DWSRF appropriations, 0.33% for financial assistance to several Trusts and Territories, $10 million for health effects research on drinking water contaminants, $2 million for the costs of monitoring for unregulated contaminants, and up to 2% for technical assistance. The EPA may use 1.5% of funds each year for making grants to Indian Tribes and Alaska Native villages.[8]

DRINKING WATER SECURITY

The 107th Congress passed the Public Health Security and Bioterrorism Preparedness and Response Act of 2002 (P.L. 107-188) to address a wide range of security issues. Title IV of the Bioterrorism Act amended SDWA to address threats to drinking water security. Key provisions are summarized below.

Vulnerability Assessments

Under new SDWA Section 1433, each community water system serving more than 3,300 people was required to conduct an assessment of the system's vulnerability to terrorist attacks or other intentional acts intended to disrupt the provision of a safe and reliable drinking water supply. This provision established deadlines, based on system size, for community water systems to certify to the EPA that they had conducted a vulnerability assessment and to submit to the EPA a copy of the assessment. Section 1433 exempted the contents of the vulnerability assessments from disclosure under the Freedom of Information Act (except for information contained in the certification identifying the system and the date of the certification), and it provides for civil and criminal penalties for inappropriate disclosure of information by government officials.

Section 1433 further required each of these community water systems to prepare an emergency response plan incorporating the results of the vulnerability assessment. The EPA was directed to provide guidance to smaller systems on how to conduct vulnerability assessments, prepare emergency response plans, and address threats.

Section 1433(e) authorized the appropriation of $160 million for FY2002, and such sums as may be necessary for FY2003 through FY2005, to provide financial assistance to community water systems to conduct vulnerability assessments, to prepare response plans, and for expenses and contracts to address basic security enhancements and significant threats.

The Bioterrorism Act also added Sections 1434 and 1435 to SDWA, directing the Administrator to review methods by which terrorists or others could contaminate or otherwise disrupt the provision of safe water supplies. These provisions require the EPA to review methods for preventing, detecting, and responding to such disruptions, and methods for providing alternative drinking water supplies if a water system was destroyed or impaired. Section 1435(e) authorized $15 million for FY2002, and such sums as may be necessary for FY2003 through FY2005 to carry out Sections 1434 and 1435.

Emergency Powers

Under Section 1431, the Administrator has emergency powers to issue orders and commence civil action if (1) a contaminant likely to enter a public drinking water supply system poses a substantial threat to public health, and (2) state or local officials have not taken adequate action. The Bioterrorism Act amended this section to specify that EPA's emergency powers include the authority to act when there is a threatened or potential terrorist attack or other intentional act to disrupt the provision of safe drinking water or to impact the safety of a community's drinking water supply.

Tampering with Public Water Systems

Section 1432 provides for civil and criminal penalties against any person who tampers, attempts to tamper, or makes a threat to tamper with a public water system. Amendments made by the Bioterrorism Act increased criminal and civil penalties for tampering, attempting to tamper, or making threats to tamper with public water supplies. The maximum prison sentence for tampering increased from 5 to 20 years. The maximum prison sentence for attempting to tamper, or making threats to tamper, increased from 3 to 10 years. The maximum fine that may be imposed for tampering increased from $50,000 to $1 million. The maximum fine for attempting to tamper, or threatening to tamper, increased from $20,000 to $100,000.

Emergency Assistance

SDWA Subsection 1442(b) authorizes the EPA to provide technical assistance and to make grants to states and public water systems to assist in responding to and alleviating emergency situations. The Bioterrorism Act amended Subsection 1442(d) to authorize appropriations for such emergency assistance of not more than $35 million for FY2002, and such sums as may be necessary for each fiscal year thereafter.

ADDITIONAL SDWA PROVISIONS

Lead-Free Plumbing

Section 1417 broadly prohibits the use of any pipe, any pipe or plumbing fitting or fixture, solder, or flux in the installation or repair of public water systems or plumbing in residential or nonresidential facilities providing drinking water that is not "lead free" (as defined in the act). This section also makes it unlawful to sell solder or flux that is not lead-free (unless it is properly labeled), or pipes, plumbing fittings, or fixtures that are not lead-free, with the exception of pipes used in manufacturing or industrial processing or other specific exceptions.[9] Added in 1986, Section 1417(d) defined "lead free" to mean not more than 0.2% lead for solders and fluxes, and not more than 8% lead for pipes and pipe fittings. In 1996, Congress added Section 1417(e), directing EPA to issue regulations setting health-based performance standards limiting the amount of lead that may leach from new plumbing fittings and fixtures unless a voluntary industry standard was established within one year of enactment. An industry standard was established.

Enacted January 4, 2011, the Reduction of Lead in Drinking Water Act (P.L. 111-380) amended Section 1417 to revise the SDWA definition of "lead free" and to add new exemptions from prohibitions on the use or sale of lead pipes, plumbing, and fittings and fixtures. The act reduced the allowable level of lead in products in contact with drinking water from 8.0% to 0.25% (weighted average),[10] and exempted from the general prohibitions "(A) pipes, pipe fittings, plumbing fittings, or fixtures, including backflow preventers, that are *used exclusively for nonpotable services* such as ... industrial processing, irrigation, outdoor watering or any other uses where the water is not anticipated to be used for human consumption"; and (B) various specified products including tub fillers, shower valves, service saddles, or water distribution main gate valves at least 2 inches in diameter. (Italics added.)

P.L. 111-380 removed the reference to Section 1417(e), which required that plumbing fittings and fixtures "intended by the manufacturer to dispense water for human ingestion" must comply with the industry standard; rather, these products became subject to the definition of "lead free" in Section 1417(d). The provisions of P.L. 111-380 became effective on January 4, 2014, and any product that does not meet the 0.25% lead limit may no longer be sold or installed, unless exempt from the general prohibitions.

In October 2013, EPA announced that fire hydrants would not qualify for the exclusion for pipes, fittings, and fixtures used exclusively for non-potable services and, thus, would be required to meet the new lead-free standards.[11] In December 2013, the House and Senate passed H.R. 3588, the Community Fire Safety Act of 2013, to amend the SDWA to clearly exempt fire hydrants from coverage under Section 1417. The President signed the bill into law (P.L. 113-64) on December 20, 2013. P.L. 113-64 also directs EPA to (1) consult with the National Drinking Water Advisory Council on potential revisions to the SDWA regulations for lead, and (2) request the Council to consider lead sources throughout the drinking water distribution system, including components used to reroute water during repairs.[12]

Research, Technical Assistance, and Training

Section 1442 authorizes the EPA to conduct research, studies, and demonstrations related to the causes, treatment, control, and prevention of diseases resulting from contaminants in water. The agency is directed to provide technical assistance to the states and municipalities in administering their public water system regulatory responsibilities. This section authorized $15 million annually for technical assistance to small systems and Indian Tribes, and $25 million for health effects research. (Title II of P.L. 104-182, the 1996 amendments, authorized additional appropriations for drinking water research, not to exceed $26.6 million annually for FY1997 through FY2003.)

Demonstration Grants

The Administrator may make grants to develop and demonstrate new technologies for providing safe drinking water and to investigate health implications involved in the reclamation/reuse of waste waters (§1444).

Records, Inspections, and Monitoring

Section 1445 states that persons subject to requirements under SDWA must establish and maintain records, conduct water monitoring, and provide any information that the Administrator may require by regulation to carry out the requirements of the act. Section 1445(b) authorizes the Administrator or a representative, after notifying the state in writing, to enter and inspect the property of water suppliers or other persons subject to the act's requirements, to determine whether the person is in compliance with the act. Failure to comply with these provisions may result in civil penalties.

This section also directs the EPA to promulgate regulations establishing the criteria for a monitoring program for unregulated contaminants. Beginning in 1999 and every five years thereafter, the EPA must issue a list of not more than 30 unregulated contaminants to be monitored by public water systems. States are permitted to develop representative monitoring plan to assess the occurrence of unregulated contaminants in small systems, and the section authorized $10 million to be appropriated for each of FY1999 through FY2003 to provide grants to cover the costs of monitoring for small systems. All monitoring results are to be included in a national drinking water occurrence data base created under the 1996 amendments.

National Drinking Water Advisory Council

The act established a National Drinking Water Advisory Council, composed of 15 members (with at least two representing rural systems), to advise, consult, and make recommendations to the Administrator on activities and policies derived from the act (§1446).

Federal Agencies

Any federal agency having jurisdiction over federally owned public water systems must comply with all federal, state and local drinking water requirements as well as any underground injection control programs. The act provides for waivers in the interest of national security (§1447).

Assistance to Colonias[13]

Added in 1996, Section 1456 authorized the EPA and other appropriate federal agencies to award grants to Arizona, California, New Mexico, and Texas to provide assistance (not more than 50% of project costs) to *colonias* where the residents are subject to a significant health risk attributable to the lack of access to an adequate and affordable drinking water system. Congress authorized appropriations of $25 million for each of fiscal years 1997 through 1999.

Estrogenic Substances

Section 1547 authorized the EPA to use the estrogenic substances screening program created in the Food Quality Protection Act of 1996 (P.L. 104-170) to provide for testing of substances that may be found in drinking water if the Administrator determines that a substantial population may be exposed to such substances.

Drinking Water Studies

Section 1458 directed the EPA to conduct drinking water studies involving subpopulations at greater risk and biological mechanisms. EPA also was directed to conduct studies to support specific regulations, including those for disinfectants and disinfection byproducts and *Cryptosporidium.*

Selected P.L. 104-182 Provisions Not Amending SDWA

The 104th Congress included a variety of drinking water-related provisions in the 1996 SDWA Amendments that did not amend the Safe Drinking Water Act. Several of these provisions are described below.

Transfer of Funds

Section 302 authorized states to transfer as much as 33% of their annual drinking water state revolving fund grant to the Clean Water Act (CWA) SRF, or an equivalent amount from the CWA SRF to the DWSRF through FY2001.

In several subsequent conference reports for EPA appropriations, Congress authorized states to continue making transfers between the two funds, and in P.L. 109-54, Congress made this authority permanent.[14]

Grants to Alaska

Section 303 of the 1996 amendments authorized the EPA to make grants to the State of Alaska to pay 50% of the costs of improving sanitation for rural and Alaska Native villages. Grants are for construction of public water and wastewater systems, and for training and technical assistance programs. Appropriations were authorized at $15 million for each of fiscal years 1997 through 2000. (In P.L. 106-457, Congress reauthorized appropriations for these rural sanitation grants at a level of $40 million for each of fiscal years 2001 through 2005.)

Bottled Water

Section 305 revised Section 410 of the Federal Food, Drug, and Cosmetic Act to require the Secretary of Health and Human Services to issue bottled drinking water standards for contaminants regulated under SDWA within 180 days after the EPA promulgates the new standards, unless the Secretary determines that a standard is not necessary.

Wastewater Assistance to Colonias

Section 307 authorized EPA to make grants to colonias for wastewater treatment works. Appropriations were authorized at $25 million for each of fiscal years 1997 through 1999.

Additional Infrastructure Funding

Section 401 authorized additional assistance, up to $50 million for each of fiscal years 1997 through 2003, for a grant program for infrastructure and watershed protection projects.

Table 2. U.S. Code Sections of the Safe Drinking Water Act (Title XIV of the Public Health Service Act) (42 U.S.C. 300f-300j-26)

42 U.S.C.	Section Title	SDWA(as amended)
Chapter 6A	Public Health Service	
Subchapter XII	*Safety of Public Drinking Water Systems*	
Part A	Definitions	
300f	Definitions	§1401
Part B	Public Water Systems	
300g	Coverage	§1411
300g-1	National drinking water regulations	§1412
300g-2	State primary enforcement responsibility	§1413
300g-3	Enforcement of drinking water regulations	§1414
300g-4	Variances	§1415
300g-5	Exemptions	§1416
300g-6	Prohibitions on the use of lead pipes, solder, and flux	§1417
300g-7	Monitoring of contaminants	§1418
300g-8a	Operator certification	§1419
300g-9a	Capacity development	§1420
Part C	Protection of Underground Sources of Drinking Water	
300h	Regulations for state programs	§1421
300h-1	State primary enforcement responsibility	§1422
300h-2	Enforcement of program	§1423
300h-3	Interim regulation of underground injections	§1424
300h-4	Optional demonstration by states relating to oil and natural gas	§1425
300h-5	Regulation of state programs	§1426
300h-6a	Sole source aquifer demonstration program	§1427
300h-7a	State programs to establish wellhead protection areas	§1428
300h-8a	State ground water protection grants	§1429
Part D	Emergency Powers	
300i	Emergency powers	§1431
300i-1	Tampering with public water systems	§1432
300i-2a	Terrorist and other intentional acts	§1433
300i-3	Contaminant prevention, detection, and response	§1434
300i-4[a]	Supply disruption prevention, detection and response	§1435
Part E	General Provisions	
300j	Assurance of availability of adequate supplies of chemicals necessary for treatment of water	§1441
300j-1[a]	Research, technical assistance, information, training of personnel	§1442
300j-2[a]	Grants for state programs	§1443
300j-3[a]	Special project grants and guaranteed loans	§1444
300j-4[a]	Records and inspections	§1445
300j-5	National Drinking Water Advisory Council	§1446
300j-6	Federal agencies	§1447
300j-7	Judicial reviews	§1448
300j-8	Citizen civil actions	§1449
300j-9	General provisions	§1450
300j-11	Indian Tribes	§1451
300j-12[a]	State revolving loan funds	§1452
300j-13	Source water quality assessment	§1453
300j-14[a]	Source water petition program	§1454

Table 2. (Continued)

42 U.S.C.	Section Title	SDWA(as amended)
300j-15	Water conservation plan	§1455
300j-16[a]	Assistance to colonias	§1456
300j-17	Estrogenic substances screening program	§1457
300j-18[a]	Drinking water studies	§1458
Part F	Additional Requirements to Regulate the Safety of Drinking Water	
300j-21	Definitions	§1461
300j-22	Recall of drinking water coolers with lead-lined tanks	§1462
300j-23	Drinking water coolers containing lead	§1463
300j-24	Lead contamination in school drinking water	§1464
300j-25[a]	Federal assistance for state programs regarding lead contamination in school drinking water	§1465
300j-26	Certification of testing laboratories	[b]

Note: This table shows only the major code sections. For more detail and to determine when a section was added, the reader should consult the official printed version of the U.S. Code.

[a] These sections include authorizations of appropriations.

[b] This provision was added by the Lead Contamination Control Act (P.L. 100-572, §4), which did not amend SDWA.

End Notes

1 EPA's long-standing policy is to exclude transient systems from drinking water regulations except for those contaminants, such as nitrate, that the EPA believes have the potential to cause immediate adverse human health effects resulting from short-term exposure. (Source: Environmental Protection Agency. National Primary Drinking Water Regulation on Lead and Copper, minor revisions. January 12, 2000 (65 Federal Register 1950).)

2 For a discussion of EPA regulation of contaminants under the revised process, see, for example, U.S. Government Accountability Office, EPA Should Improve Implementation of Requirements on Whether to Regulate Additional Contaminants, GAO-11-254, May 27, 2011, 146 p. http://www.gao.gov/products/GAO-11-254.

3 For enforcement information and statistics, see U.S. Environmental Protection Agency, Providing Safe Drinking Water in America: 2011 National Public Water Systems Compliance Report, Office of Enforcement and Compliance Assurance, EPA document No. 305R13002. May 16, 2011, 94 p. http://www.epa.gov/ compliance/resources/reports/accomplishments/sdwa/sdwacom2011.pdf.

4 According to the EPA, of roughly 156,600 public water systems, 142,400 rely on ground water and 14,200 rely on surface water. Among roughly 51,700 community water systems, 40,000 rely on ground water and 11,700 rely on surface water.

5 The Energy Policy Act of 2005 (P.L. 109-58, §322) amended SDWA Section 1421(d) to specify that the definition of "underground injection" excludes the injection of fluids or propping agents (other than diesel fuels) used in hydraulic fracturing operations related to oil, gas, or geothermal production activities.

6 The wells are classified as follows: Class I (inject hazardous wastes, industrial non-hazardous liquids, or municipal wastewater beneath the lowermost USDW); Class II (inject brines and other fluids associated with oil and gas production, and hydrocarbons for storage); Class III (inject fluids associated with solution mining of minerals beneath the lowermost USDW); Class IV (inject hazardous or radioactive wastes into or above USDWs and generally are banned); Class V (all injection wells not covered under other classes—many of these wells inject non-hazardous fluids into or above USDWs, and are typically shallow, on-site disposal systems), and Class VI (inject carbon dioxide (CO2) for long-term geologic sequestration to reduce atmospheric emissions of CO2 from industrial sources).

[7] See U.S. Environmental Protection Agency, Drinking Water Infrastructure Needs Survey and Assessment: Fifth Report to Congress, Office of Water, EPA 816-R-13-006, April 2013, 70 p., http://water.epa.gov/grants_funding/dwsrf/ upload/epa816r13006.pdf.

[8] For more information, CRS Report RS22037, Drinking Water State Revolving Fund (DWSRF): Program Overview and Issues, by Mary Tiemann.

[9] 42 U.S.C. 300g-6. From 1986 through January 3, 2014, "lead free" under SDWA Section 1417(d) was defined to mean not more than 0.2% lead for solders and fluxes, and not more than 8% lead for pipes and pipe fittings. The 1996 SDWA amendments further provided that, for plumbing fittings and fixtures, "lead free" referred to plumbing fittings and fixtures in compliance with industry standards established under Section 1417(e).

[10] As amended in 2011, Section 1417(d) defines "lead free" to mean "(A) not containing more than 0.2% lead when used with respect to solder and flux (unchanged from existing law); and (2) not more than a weighted average of 0.25% lead when used with respect to the wetted surfaces of pipes, pipe fittings, plumbing fittings, and fixtures." P.L. 111-380 established a formula to calculate the weighted average lead content of a pipe, pipe fitting, plumbing fitting, or fixture.

[11] U.S. Environmental Protection Agency, Summary of the Reduction of Lead in Drinking Water Act and Frequently Asked Questions, EPA 815-S-13-001, October 2013, p. 5.

[12] For more information, see CRS Report INB00006, Lead-Free Requirements for Fire Hydrants, by Mary Tiemann.

[13] Colonias generally are described as unincorporated communities or housing developments on the U.S. side of the U.S.-Mexico border that lack some or all basic infrastructure including plumbing and public water and sewer.

[14] The Department of the Interior, Environment, and Related Agencies Appropriations Act, 2006, P.L. 109-54, Title II, August 2, 2005, 119 Stat. 530, provided "That for fiscal year 2006 and thereafter, State authority under section 302(a) of P.L. 104-182 shall remain in effect."

INDEX

D

H

I

J

K

L

M

N

O

P

Q

R

S

T

U

V

W

Y